The Smart Guide to
Geothermal

How to Harvest Earth's Free Energy
for Heating & Cooling

Donal Blaise Lloyd

Foreword by Michael Hunt

PIXYJACK PRESS INC

The Smart Guide to Geothermal:
How to Harvest Earth's Free Energy for Heating & Cooling

Copyright © 2011 by Donal Blaise Lloyd

No part of this book may be reproduced, stored in a retrieval system or transmitted in any form, or by any means, electronic, mechanical, photocopying, recording or otherwise, without prior written permission of the publisher, except by a reviewer, who may quote brief passages in review.

Published by PixyJack Press, Inc.
PO Box 149, Masonville, CO 80541 USA

First Edition 2011

9 8 7 6 5 4 3 2 1

ISBN 978-0-9773724-8-5

Library of Congress Cataloging-in-Publication Data

Lloyd, Donal Blaise.
  The smart guide to geothermal : how to harvest Earth's free energy for heating & cooling / Donal Blaise Lloyd ; foreword by Michael Hunt. -- 1st ed.
    p. cm.
  Includes bibliographical references and index.
  Summary: "Covers residential geothermal heating and cooling, including various system options and installation configurations, costs and payback issues, performance standards, and contractors. Also examines how energy-efficient, non-polluting geothermal heat pumps work and how to integrate solar energy"-- Provided by publisher.
  ISBN 978-0-9773724-8-5
  1. Ground source heat pump systems. I. Title.

TH7417.5.L64 2011
697--dc22   2010051015

 Printed in Canada
on chlorine-free, 100% postconsumer recycled paper

Geothermal illustrations by Will Suckow.
Book design by LaVonne Ewing.

To Alan and Lydia who took over the stewardship of our 1930 schoolhouse and were the first to introduce into my head the concept of geothermal heat pumps. That initiated a chain of events, including these pages, which is still reverberating.

Contents

Foreword	*ix*
Preface	*xi*
Introduction: It Is More Efficient to Transfer than Burn	*15*

Part I — The Big Picture: What Geothermal Energy Is and Why You Want It

1	A Different Approach to Heating & Cooling	*29*
2	How I Learned to Get Even with the Oil Company	*33*
3	How a Geothermal Heat Pump (GHP) Works: Borrowing Heat from the Earth	*41*
4	Ground-Source Solutions: Open vs. Closed Loops	*50*
5	Direct Exchange (DX) Geothermal Systems	*62*
6	Using Your Well as the Heat Source	*67*
7	Geothermal Advantages/Disadvantages	*71*
8	Does a Geothermal System Really Increase a Home's Market Value?	*74*
9	Learning to Live With a GHP System	*77*
10	Reducing Operating (Electrical) Costs	*83*
11	Geothermal Plus Solar Energy: A Perfect Match	*89*
12	What Will It Cost?	*95*
13	Taking Advantage of Rebates & Tax Credits	*101*
14	Payback: When Do You Get Your Money Back?	*105*
15	Finding a Contractor	*111*

continued

Part II — Unraveling the Science & Technology of Geothermal Heat Pumps

16	The Inside Story of Geothermal Super Efficiencies	*121*
17	A Look Inside: How GHPs Deliver Heat Into Your Home	*131*
18	The Importance of GHP Performance Standards	*136*

Part III — The Broader View: A Look Beyond Residential to the Global Market

19	The Geothermal Marketplace	*143*
20	Commercial/Institutional Geothermal Systems	*146*
21	Future Trends	*153*
22	Decision Time & My Ultimate Geothermal Package	*157*

Earth-Smart Home Case Studies   *39, 55, 58, 60, 66, 93, 118*

Epilogue: The Disappearing Stars   *160*

Appendix

A	Frequently Asked Questions	*164*
B	Helpful Links and Resources	*169*
C	Residential GHP Manufacturers	*172*
D	State Energy Offices	*177*
E	Geothermal Careers: Joining the Industry	*181*

Index	*185*
List of Graphics	*189*
Acknowledgments	*190*
About the Author & Technical Advisor Michael Hunt	*192*

Foreword

I am a GEO Junky. I started in the geothermal business back when there was no geo in the west-central United States. First, as a farmer wanting self-sustainable living from the land, I installed a geothermal system in my own home. Then as a contractor helping others obtain the same goal. After 20 years of making mistakes and experimenting to make systems work at peak performance, I attended North Dakota State University for an engineering degree. You might ask, "Why the engineering degree?" As a contractor and farmer with no formal education in HVAC or geothermal, there were many questions that I could not find answers for. College provided those answers and sparked an interest in what lies below the earth's surface and how we could capture that vast energy. Yes, oil and gas come from the earth, but I did not want to destroy that energy resource; I wanted to *move* energy. I wanted to learn how to harness the energy withing the earth in the most efficient way.

Taking the next step, I wanted to help educate others on how they can become more self-sufficient. And even though I work as an instructor certifying installers, I had considered writing a book. Lack of time kept me from putting pen to paper, but after meeting Don Lloyd, I felt that I could fulfill that goal by contributing to this book. If I was to write my own book, this is the book I would've written. It's not a how-to manual for installing a geothermal system, but a book that contains everything a homeowner needs to know to make an informed decision on geothermal for his/her home.

And it is critical that the homeowner understands the basic facts. With all the various forms of geothermal systems available and the multitude of variables of each home site, I've not yet seen a cookie-cutter approach. One size does not fit all.

One of the most important lessons is that quality comes with a price; cheap does not last. A perfect example is that of a building contractor who went for the low bid from an HVAC subcontractor. Six months later, the homeowners felt the heat pump was not performing well and there was a noise between the main floor and the sheet rock ceiling of their basement. As the manufacturer of their GeoFurnace heat pump, I was called in because the HVAC contractor wouldn't make the service call. After cutting holes in their basement ceiling, we discovered that the hanging material used for holding the ductwork in place had failed and parts were lying on the sheetrock, blowing back and forth as the heat pump ran. Lack of proper fasteners along with the overall poor quality of installation was the issue. The home-building contractor learned his lesson, at the homeowner's expense.

By studying the information found in the pages of this book, you—the consumer—will not become a victim of a poor geothermal installation. Just as farmers envision themselves as caretakers of the earth, so are you. Why burn, waste and pollute when a far superior heating and cooling method is available? The coming generations will thank us for our foresight.

A final note: although this book is written primarily for homeowners, geothermal industry professionals (installers, trainers and designers) will benefit from a number of chapters, especially *Part II – Unraveling the Technology and Science*. Builders and architects will also find this book a handy tool for themselves and their clients.

Respectfully,
Michael Hunt, President
GeoFurnace Manufacturing

Preface

The next generation of home heating/cooling is here! I can even foresee traditional oil burners and gas boilers becoming nearly obsolete in the future. I'm pleased to say there is an excellent alternative to fossil-fuel burning, home-heating systems and it is the geothermal heat pump.

The primary purpose of this book is to spread the word that if you are planning the construction of a new home, or have an older, aging oil or gas burner, you would be wise to consider a ground-source geothermal system. But even homeowners who have fairly new oil or gas systems are retrofitting with geothermal heat pumps, due in part to the rising price of fuel and the possibility of having zero fuel costs in the future as well as the increased value of their homes. Builders and architects should also consider how this technology can positively impact their projects. With substantial incentives available (thanks to the American Recovery and Reinvestment Act of 2009), now is the time to implement this smarter approach.

The immediate driving force behind geothermal is saving money, year after year. Added to that is the benefit of reducing our need for foreign and domestic oil, and the reduction of pollutants in the air.

As a homeowner or small business owner, your first step should be to do your homework to learn more. You have in your hands right here a survival kit; a complete picture of everything you need to know about ground-based geothermal systems. My goal is to make you a well-informed homeowner on the subject of geothermal heat pumps (GHP) and to encourage you to implement this technology in your own home.

The key to a successful geothermal system is in the installation.

There are many certified, qualified and experienced contractors out there and new installers are continually coming into the field, but their supply is not keeping up with the industry's growth. Official US standards exist for the *hardware* in terms of quality, safety and performance, but there are no officially approved national Air-Conditioning, Heating and Refrigeration Institute (AHRI) *installation* standards as of 2010. In the interim, the International Ground Source Heat Pump Association (IGSHPA) provides an installation standard that is used for accreditation training of installers in the United States. Canada already has C448 in place as an official installation standard. This puts the responsibility on the homeowners to be knowledgeable about this technology so that they are in a more comfortable position to select a competent contractor/installer and to work with them as discussed in *Chapter 15: Finding a Contractor.* It is not an overwhelming task for a homeowner, yet it is very important.

Installing a geothermal system in my new home was one of the smartest decisions I ever made! This book will make you far better equipped to make your own decision. It is organized into three parts:

Part I—The Big Picture is a general description of geothermal heat pump systems (including my personal experience), the advantages, disadvantages, costs, payback time and increased market value of the home. This section also covers finding a contractor, gives examples of homes with installed geothermal systems, living with a GHP system, and much more.

Part II—Unraveling the Science and Technology goes into slightly more technical depth in describing the flow of heat from the ground to the house, the "magic" of the compressor, how 400–500% efficiency is obtained, and how the Laws of Thermodynamics come into play. Homeowners, GHP manufacturers and installers will especially welcome chapters 16 and 17 that go deep inside the heat pump to clearly explain why it works.

Part III—The Broader View portrays a broader picture of the geothermal world and discusses future trends. It also includes *The Ultimate Geothermal Package* where I go out on a limb to tie together all of the information in the book and present my idea of what I believe to be the ideal ground-based geothermal heat pump solution and why.

The appendix includes references and links to other geothermal websites and state energy offices, plus a listing of companies that manufacture ground-based residential geothermal heat pumps in the United States and Canada. For those interested in joining the geothermal industry, a career section is also included.

This book starts out locally (a few feet below the ground in our backyard) and ends with a global outlook—all with the same goals of lowering heating costs, reducing pollution with renewable sources, and creating energy independence.

I would like to thank Michael Hunt, a long-time professional in this geothermal business, for contributing his knowledge and experience. I think you'll be pleased with the result—an unusual synergism that provides an ideal basis for making your own decision about geothermal.

Don Lloyd
January 2011

Introduction

It is More Efficient to Transfer Than Burn

I was reading a book recently about a family having a dinner conversation. One of the young daughters was discussing a history-class subject where in past times people had killed whales in order to provide oil for burning in lamps. "What a terrible waste!" she said. I think that our yet unborn great-grandchildren will look back on our times and also say, "They actually *burned* oil? What a terrible waste!" All generations do what they must in order to procure the energy they need, whether it is whale oil, coal, gas or petroleum, but there is now a smarter way, an alternative.

    The next generation of heating systems using proven refrigerator technology is readily available for homeowners—it is called ground-source, geo-exchange, earth-coupled, or geothermal heat pump (GHP). This technology has been slowly perfected over the

....

According to the Environmental Protection Agency, geothermal ground-source heat pump systems are one of the most energy efficient, environmentally clean, and cost-effective space conditioning systems available.
....

last 50 years to the point where GHPs are now slowly becoming mainstream technology for domestic heating and cooling, and indeed for commercial and institutional applications.

It is far more efficient, less costly and makes more sense to *transfer* existing heat energy from a source that is renewable: the earth. It is not necessary to go deep to capture heat from the molten center of the earth. GHPs capture heat energy under our feet in the backyard, just below the frost line, and that heat energy is continually renewed by the sun as well as the earth's center. As long as the sun shines (still 5 billion years to go!), it will pump heat energy year-round into the ground beneath our feet. And we can tap into that easily, right now.

My research was conducted to make certain that this book contains not only everything you need to know about ground-based geothermal heat pumps, but also a lot more that you may be curious about. Embedded in the sum of all these chapters is one single concept that is a *must* for you to understand. It is the central reason for a geothermal heat pump to even exist. It is the reason they are being installed in increasing numbers worldwide. It is the reason subsidies and tax credits are being provided in the United States and Canada. It is the reason I installed such a system in my own home...

A GHP is superior to any other method of home heating and cooling in three very important ways:

1) Does not combust any fossil fuels
2) Uses a renewable resource
3) Has the lowest cost of operation

Figure A-1 on the next page puts that all together.

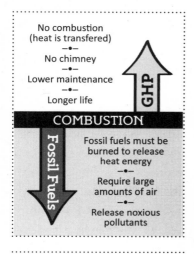
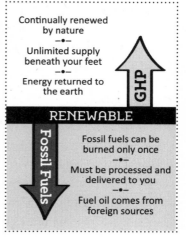
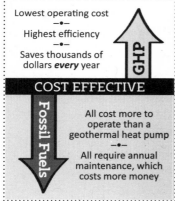

Figure A-1

In all cases, a geothermal heating/cooling system comes out on top.

GHP = Geothermal Heat Pump
Fossil Fuels = fuel oil, natural gas, propane, coal

Combustion

Let's face it; we have been heating our homes with almost Stone Age technology. Burning fossil fuels (using combustion) means flames, and much of the heat energy from flames is wasted. At the same time, combustion produces pollutants, including deadly carbon monoxide.

Fuel oil, coal, natural gas, propane, cord wood, and pellets must all be burned to release their heat energy. Burning also takes large amounts of oxygen out of your home and requires a chimney for venting the exhaust (except a few non-vent propane heaters). Electric baseboard heating does not require burning within the home, but we must remember that a significant amount of electricity used in the United States is generated by coal burning plants—plants that are only 35–45% efficient, many of which are still equipped with towering, belching chimneys.

In contrast, a GHP system produces not a single spark. It offers emission-free operation on site. Heat energy is transferred; nothing is burned. No chimney is required. Heat pumps are among the few reliable and widely available heating technologies that can deliver thermal comfort with zero emissions. The IEA (International Energy Agency) Heat Pump Centre has stated that heat pumps are one of the most significant available technologies that can offer large carbon dioxide (CO_2) reductions. In addition, GHPs require no on-site fuel storage. With less system stress, a GHP has a much longer life and requires less maintenance. (*Chapter 9 – Learning to Live With a GHP System* also provides actual experience in this area.)

That is one star for a GHP — ✪

Renewable Resource

You can only burn fossils fuel once, but a GHP system transfers heat from the earth; heat that is forever renewed by the sun and the earth's center. Fuel oil, coal, natural gas (NG), and propane (LP): none of these fuels are renewable. All are fossil fuels, and when burned, all release pollutants into the air when they are burned. Figure A-2 shows the percent of housing units by heat type from US Census data. No GHP in the census list. Maybe someday we can change that.

Type of Heating in Occupied Housing Units

Figure A-2

"Other" refers to coal, wood, solar and none.

Source: US Energy Information Administration, 2007

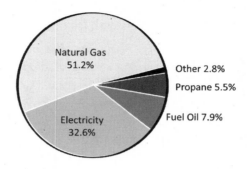

Propane, also known as LP gas or LPG, is derived from petroleum. It is not as clean as natural gas, but it can be delivered by truck as a liquid under pressure; no extensive underground pipe infrastructure is required. The National Propane Gas Association states that in 2008 LPG heated 8.1 million US homes.

Natural gas (NG) is 70% methane plus impurities such as sulfur. It is shipped in tankers and stored as LNG, liquefied natural gas. NG is the cleanest of the fossil fuels but still releases CO_2, nitrogen oxides and sulfur oxides. The Census Bureau (2007) states that over 56.7 million US homes use NG (51% of all homes).

Fuel oil: about 30% of our #2 fuel oil is produced in the United States; the remaining 70% must be imported as crude oil. The Census Bureau (2007) states that 9.3 million US homes (8% of all homes) heat with fuel oil, mostly in the northeast.

Coal, a readily combustible sedimentary rock consisting primarily of carbon, emits carbon dioxide emissions during burning that are double that of NG. It is by far the most abundant fuel produced in the United States. Anthracite coal, the hardest type, is used for home heating and it is cheap. Coal burning, self-stoking stoves are available, but they do create ash and dust. Coal is a messy fuel, but can be an option, especially if you live

> ### Homes with Oil Burning Furnaces
>
> I must confess that the low number of oil users came as a great surprise to me. In part because of corporate moves, my wife and I have owned homes in Maryland, Minnesota, Massachusetts and New York State. Every one of those homes had oil burners. I had a mind-set that most homes, except those of city dwellers, burned oil. But that is not true.
>
> Home heating sources have changed over the past 50 years. In 1960, 32.5% of US households heated with fuel oil, but only 7.9% in 2007. On the other hand, electricity was used for heat in only 1.8% of homes in 1960, but 32.6% used electricity in 2007. Propane usage stayed about the same, but natural gas heating has grown from 43.1% to 51.2%. Coal dropped from 12.2% to 0.1%, thankfully. SOURCE: *U.S. Energy Information Administration*

near the eastern Pennsylvania anthracite area. However, many coal-burning customers in that area are finding that in spite of having 7 billion tons of anthracite coal underground, mines are closing down, causing shortages.

As far as the number of homes heated by coal, the US Energy Information Administration (EIA) says that 90,000 homes were heated with coal in 2007 (down from nearly 6.5 million in 1960). However, in the past few years sales of coal stoves have increased about 10% per year.

Wood (pellets and cordwood): All are renewable, of course. Most homes use them as auxiliary or backup heat. In 1960, 4.2% of US households heated with wood, but that number has dropped to 1.3% in 2007 (US EIA).

Electric baseboard: Sources for our electricity include natural gas, coal, oil, nuclear and renewable energy *(see Figure A-3)*, but even these sources vary greatly from region to region, utility to utility. Today approximately 36.1 million homes are heated by electricity, one third of the total. It is interesting to note that 60% of these are in the South.

In summary: It requires a small amount of electrical energy to transfer a large amount of heat energy in a geothermal system. How this takes place is described in *Chapter 3 – How it Works*. So far, that is another star, adding up to two for a GHP — ✪✪

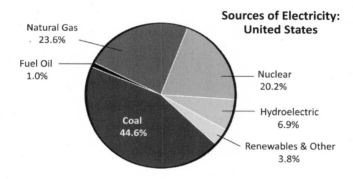

"Renewables" includes biomass, geothermal, solar and wind.
Source: US Energy Information Administration, Annual Energy Review 2009

Figure A-3

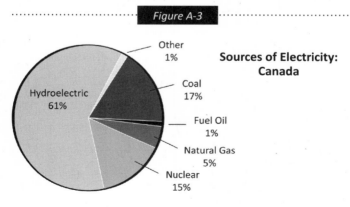

Source: Canadian Office of Energy Efficiency, 2008

Lowest Cost of Operation

You have heard the expression, "Follow the Money." Well this is where the money is. The cost of operation for a heating system can mean a difference of thousands of dollars per year—*every year*. And the factor that most heavily influences that cost is the efficiency of converting the fuel type into heat energy.

Every fuel source has a different basis or unit—gallon, ton, kilowatt-hour (kWh), etc. The only way to compare the cost of operation is to define the heat content in each of those units in terms of a common energy unit, Btu (British thermal unit) for example. This provides a dollar value per Btu (actually, it is handier to use dollars per million Btu). Finally, we must modify that cost-per-Million-Btu (MMBtu) by the actual efficiency of

Heating Cost Comparison Data

Fuel Type	Fuel Unit	Fuel Price per Unit	Btu per Unit	Fuel Price per Million Btu	Efficiency	Fuel Cost per Million Btu
GHP	kWh	$0.116	3,412	$33.90	400%	$8.48
Coal	Ton	$200.00	25,000,000	$8.00	75%	$10.67
Pellets	Ton	$250.00	16,500,000	$15.15	68%	$22.28
Wood	Cord	$200.00	22,000,000	$9.09	55%	$16.53
Propane	Gallon	$1.93	91,333	$21.16	78%	$27.13
Natural Gas	Therm*	$1.25	100,000	$12.50	78%	$16.03
Electric Baseboard	kWh	$0.116	3,412	$33.90	100%	$33.90
#2 Fuel Oil	Gallon	$2.33	138,690	$16.80	78%	$21.54

*Therm = 100,000 Btu

Figure A-4
A GHP (geothermal heat pump) has the lowest cost per BTU of any other heat source. Source: US Dept of Energy, 2008

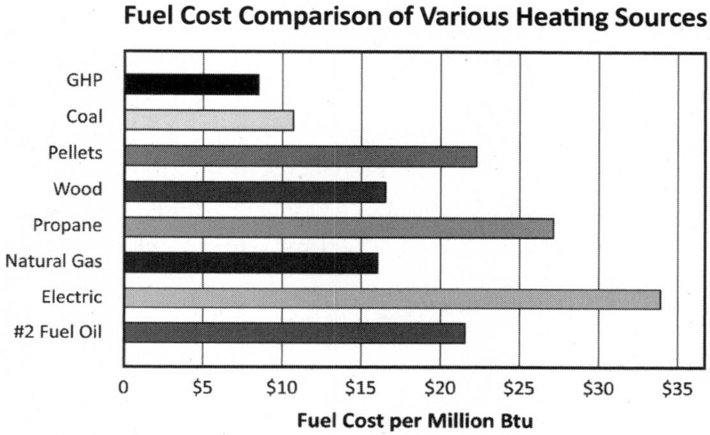

Fuel Cost Comparison of Various Heating Sources

(bar chart showing Fuel Cost per Million Btu for: GHP, Coal, Pellets, Wood, Propane, Natural Gas, Electric, #2 Fuel Oil)

Figure A-5
A GHP has an even lower cost per BTU than coal. Source: US Dept of Energy, 2008

converting a particular fuel source into heat energy. Then we have a *real* cost per MMBtu.

Figure A-4 shows how the US Department of Energy chart computed the fuel cost per MMBtu for each fuel type. Their conclusion is very clear: **A GHP has the lowest cost per energy content than any other fuel type**. This point is even more obvious when the data is converted to a bar graph as shown in Figure A-5.

This cost-savings is so important it is worth two more stars for a GHP with a grand total of four— ✪✪✪✪

To carry this one step further, we can examine the trend showing how this may change over time. I have taken DOE data from the past and present and then made some guesses about the future based on the following:

1) I assume that **oil price**s will inevitably bounce up and down but will average a small increase over the next ten years.

New crude oil production activity will not pay off until after 10 years.

2) While **natural gas** is at low demand and a relatively low price today, demand will increase as the economy improves. But this will be balanced by increased production and new sources such as oil shale. The result will probably be a future small increase in price.

3) Electric rates will continue to increase less than 5% per year, but **geothermal heat pump** efficiencies will increase even more as GHP components continue to steadily improve and as the US EPA continues to mandate increased future efficiency standards.

The results of this projection show that GHPs will have an increasing advantage over time. Two factors are involved here. One is the cost of fuel, which is assumed to be increasing for all

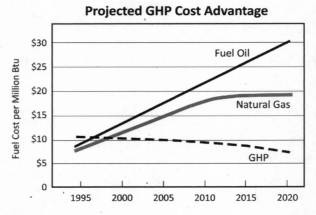

Figure A-6
Future GHP Systems will increase their cost advantage.

types. The other factor is the efficiency of converting the fuel to heat energy. In the past few years, high-efficiency oil burners and gas boilers have become available, but GHP efficiencies have not yet peaked and are still increasing. In time, newer GHP components will be able to do the same work with less electricity. That is why you see in Figure A-6 a GHP cost advantage will be even better in the future.

The trend is interesting. It shows the GHP relative cost over time compared to natural gas and heating oil. The result, based on the above assumptions, shows an increased future advantage.

So, if a GHP is so talented and the most modern system, why is it not *the* choice for every residence on the continent? The answer in part is due to lack of knowledge, as covered in more detail below. But to be fair, there are disadvantages to GHP systems. The most obvious is that the total cost to install a GHP will usually be higher than a conventional heating system because of the need for ground excavation or drilling. Even that disadvantage is not necessarily bad news. *Chapter 14 – Payback: When Do You Get Your Money Back?* includes some interesting revelations.

The Need for a GHP Book

All of this information is useless if no one is aware of it. Ground-based geothermal heat pumps are underutilized because they are not well known or are misunderstood. My architect felt that a GHP was beyond my budget. My builder had never built a home with a GHP. But the biggest problem was my bank. The bank construction-loan underwriter did not fully understand this technology. All were obstacles that had to be overcome by education and persistence.

I am not alone in this. A 2008 Oak Ridge National Laboratory study stated: "Lack of consumer knowledge and/or confidence in GHP system benefits is one of the key barriers to growth of the GHP industry." The Canadian GeoExchange Coalition (CGC) acknowledges a lack of public knowledge by including in its charter a goal of "increasing awareness."

In addition, a recent survey reported that 9 out of 10 people know what a solar panel is, while only 2 out of 10 know what a geothermal heat pump is.

From 2005 to 2008, shipments of GHPs in the United States doubled (121,243 units in 2008), in great part due to the surge in oil prices. But in 2008 alone, new housing starts in the US plus Canada totaled over 1.1 million. As you can see, GHP shipments are still a very small part of the total.

While this book has a North American perspective, the material is international in application. In fact, we have been falling behind in the application of this technology compared to Europe and Asia.

This book is a modest attempt to change that. We must find a way to capture public and political imagination. This is only one voice singing the praises of a different approach. It would indeed be cool and to the benefit to all countries if that voice swelled into a quartet, then a chorus and even into a grand orchestra. ✪

Different Terminology; Same Technology
Geothermal Heat Pump (GHP)
Ground-Source Heat Pump (GSHP)
Geoexchange (GX)
Earth-Coupled
Earth Energy

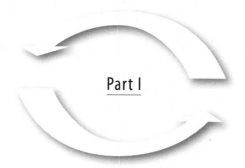

Part I

The Big Picture:
What Geothermal Energy Is and Why You Want It

Chapter 1

A Different Approach to Heating and Cooling

What are you paying to heat your home this year? Responses to that question generally run from $2,000 to $5,000 per home. The Institute of Social and Economic Research (ISER) conducted a study in Alaska. The average 2,500 square-foot (232 sq m) home paid $4,500 in 2008.

My last home, a renovated two-story brick, 8,000 square-foot (743 sq m) former schoolhouse, had an annual bill in 2005 of over $7,000 that has since grown much higher. This was the major reason for selling it, even though my wife and I had put our hearts and souls and many years of our lives into that building.

At that time, I had a very real concern about how much I would pay for fuel in the future. But now I have no uncertainty about *my* fuel costs: zero! Why?

Fuel oil and propane were two options where I could not control prices. Natural gas was not even available. But I could do without burning fossil fuels and so I took a different approach. I installed a ground-based geothermal heat pump (GHP) in our new home. Designing and building our home was a four-way collaboration between our architect, our builder, my wife and myself. But the geothermal decision was my passion; my responsibility for better or worse.

What I envisioned was a heating system that would provide comfortable heating and full-house air-conditioning (AC), did not have to burn anything, never needed the nozzles cleaned (it had no nozzles), emitted no noxious gases up the chimney (no chimney at all), provided partial free hot water, had lower operating and maintenance costs than an oil burner, was far more efficient and quiet, and finally, saved thousands of dollars every year. And that is exactly what I got! Although it is not a common situation, the installation of my particular system cost less than an oil burner plus full AC.

The US EPA states, "Geothermal heat pump systems are the most energy efficient, environmentally clean and most cost effective space conditioning systems available." But more importantly, this underutilized technology has the potential to be an important factor in reducing every country's dependence on foreign oil while at the same time reducing air pollution.

What is a geothermal heat pump (GHP)? In short, a GHP draws heat from the ground. An air-source heat pump, on the other hand, uses air as the source of heat. Air-source systems can be useful in warmer climates, but are limited in efficiency and application. When comparing air-source and ground-source heat pumps, GHPs are quieter, last longer, need little maintenance and do not depend on the temperature of the outside air.

Other geothermal heating systems use naturally heated underground water, such as hot springs or reservoirs. These direct-use geothermal systems are not included in this book.

You should also note that a GHP system is not the same as geothermal energy or geothermal electric power. Geothermal power is an electrical power generating system that includes drilling deep into the earth's crust to tap into heat that can be used to turn a turbine that generates electricity *(see below)*.

A ground-based GHP uses heat just below the frost level of the ground. This heat energy is constantly renewed by the sun and by heat from the earth to provide a relatively constant temperature. The GHP takes heat from the ground in the winter to heat a home, and in the summer, it provides cooling by putting heat back into the ground. So, we are merely borrowing heat from the earth.

This technology turned out to be a very personal thing with me. Installing a geothermal heating system in my new home instead of an oil burner was not as easy as I had imagined, but clearly well worth it! ✪

Geothermal Electric Power

There is another, but quite different geothermal effort underway that is going to have a large future impact on our energy supply. You should be aware of this because in some ways it is often confused with geothermal heat pumps. Geothermal Electric Power (GEP) is similar to the geothermal heat pump in that it takes heat energy from the earth, but there are major differences. First, it is not for residential applications; it will be closer in size to a private utility operation. Second, while a GHP transfers heat energy from the earth to warm a home (or building), the GEP converts heat energy into electrical energy. The geothermal power system captures heat energy that originates at the earth's molten center—"using the earth's furnace."

The crust of the earth is made up of huge plates that are in constant but very slow motion relative to one another. Geologic processes allow molten magma to rise up near to the surface, sometimes producing hot springs, steam, even lava and volcanoes. Some of the hot water remains trapped below, forming geothermal reservoirs. Today we drill deep wells into these geothermal reservoirs to bring up hot water or steam to the surface to drive turbines to generate electricity.

But steam or hot water is just a small part of the almost unlimited potential. Usable geothermal resources will not be limited to the "shallow" hydrothermal reservoirs at the crustal plate boundaries. Much of the world is underlain (3 to 6 miles down) by hot, dry rock—no water, but lots of heat energy. Experiments are underway in the United States and many other countries to pipe water into deep holes to create more hydrothermal resources. The process involves drilling as deep as 30,000 feet or more, pumping in water under pressure, heating it, and then using that energy to drive a turbine.

This process does not burn fuels and produces no pollution. It is easy on the land. Installations in remote locations could bring power far from the electrified population centers. Water requirements, however, may be a problem in arid areas.

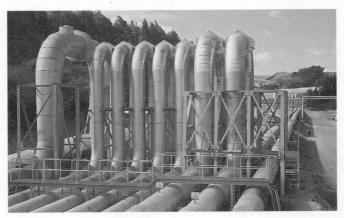

The earth is a giant heat engine—a constant cycle of heat generation, heat transfer and cooling. Geothermal power plants, like this one in New Zealand, harvest that heat and power on a very large scale.

Chapter 2

How I Learned to Get Even With the Oil Company

The Schoolhouse: How It All Started

If we had not gone to France and seen how war-damaged, 500-year-old buildings could be charmingly renovated, we probably would not have ever taken on the comparatively simpler renovation of a deteriorating 70-year-old building. And if we had not taken on the risk of renovation, we would never have discovered geothermal heat pumps.

The schoolhouse we found, almost by accident, had two stories totaling 8,000 square feet (743 sq m), a leaking roof, red brick with dying ivy, an interior vandalized after 27 years of vacancy, and crumbling plaster. It needed a new heating system, new plumbing and new electrical service. But it had solid steel-beam interior supports, arched windows, 12-foot (3.6 m) ceilings, a gymnasium/auditorium that would make a to-die-for north-light art studio, a large working septic system, a rural setting, and a certain charm and character that grabbed us.

It was not an easy decision, but we bought it and spent two years renting nearby as we lovingly restored it into a beautiful home with extensive gardens and a professional, working art studio. There was wall space for hanging very large canvases. We had fun living there, with many Boston and New York City visitors, among them art collectors, including two Boston luxury hotels, who bought my wife's work. Other artists from the area were invited to use the upstairs classrooms as art studios.

We were able to put the building on the National Register of Historic Buildings and I finally published the full story in a memoir (*Snake Mountain Trilogy: A Berkshire Memoir,* 2005). That led to a book party in the schoolhouse's former second-grade room (by then our dining room) where I told our guests, "We had our building closing on Halloween Eve, we moved in on Halloween Eve, and today is Halloween Eve. I want you to know that you are the largest, and by far the oldest trick-or-treaters we have ever had." I also explained that this was the first time I had ever publicly used my full name—Donal Blaise Lloyd. As a fourth grader, I was teased, "Blaise, Blaise, your pants are on fire!" Those kids were not only annoying; they could not even

After years of labor and love, our schoolhouse became our home, but heating costs were a real shocker.

spell. And I had a chronic problem with my first name because so many insisted on adding a "d" at the end. And when corrected, they made it worse by putting the accent on the second syllable. But I decided it was time to take charge of my name and do it my way.

But then came the bad news, really bad. The building required 3,000 gallons (11.3 kl) of fuel oil every season, but at 99 cents a gallon it was bearable. But price increases struck soon and hard! When the cost passed the $7,000 level with every indication of going higher, it was clear that we had to sell it, as difficult as that was. So we did.

The new owner, an artist who fully appreciated the artist's stewardship of the building, hired an engineering consultant to determine if a geothermal heat pump was practical considering the age, construction and size of the building. This was a new and intriguing concept. The more I learned about geothermal energy, the more hooked I became.

A Geothermal Heating System for our New Home

Starting all over again, I was determined that I was not going to be an oil captive any longer. Our new, smaller home was going to have a ground-based geothermal heating/cooling system. I did my homework to understand how GHP systems worked, their advantages and disadvantages, the types of systems available, and then studied the companies that manufacture the heat pumps. I learned about efficiencies and costs, and had to relearn the physics that makes the whole thing work.

Have you ever had the feeling that something you're considering is absolutely right? It may be a small point or a large, important item, and it does not happen often, but once in a while something clicks. It just feels right. And so it was with this endeavor for alterative heating.

We found exactly the right architect to translate our needs and wants into a beautiful, open home design with a large art studio attached. My first problem, however, came during the early design phase. I was advised by several people (including my architect) to forget geothermal—we could not afford it on our budget! Something would have to go. I knew not what, but it wasn't going to be the geothermal.

I still consider myself lucky to have found our builder. He is experienced, capable and honest, has a skilled team and, as we found out, was very flexible to our changes. He was especially good at working with us to reduce costs, but he had no experience with geothermal systems. He felt responsible for everything he built and he did not want to be responsible for this "unknown." So we made a deal. He would get bids for an oil system plus a full air-conditioning system. I would get bids from geothermal contractors.

I did my homework. I selected a manufacturer and used their website to pick an experienced local geothermal-system contractor. As he and I discussed whether we would need to dig vertical holes or horizontal trenches for the ground closed loop, I casually mentioned that I had drilled my well and that it had an excellent flow rate of 12 gallons per minute (45 lpm). He picked up on that at once and said that we should use the well as an open-loop heat source. "It will work and it will cost far less." That clinched it; the decision was a no-brainer.

The final question was what to do with the colder well water once it delivered its heat. After some engineering consulting, we decided to pump it right back into the well. This presented a potential problem. If the weather was very cold for an extended period, the well temperature could be impacted and system efficiency could be lowered. Forty degrees was the accepted minimum, so we included an automatic backup system that would temporarily transfer the water over the hill toward a stream when the water temperature dropped below 40°F (4°C). The final results: the geothermal setup ended up costing $7,000 less than an oil burner and separate AC, but I had to play the role of liaison between the builder and the contractor. So be it.

The last problem was with my bank. My application for a construction loan included every architectural drawing, building contract, bid, credit report and building permit: unbelievable detail. All banks have these unidentified, backroom underwriters with god-like powers. The final edict came down: everything was approved except two items: First, the steel roof was approved but the steel siding was not. They felt that it could act as a chimney in case of a fire. Well, that was a minor point; there were other materials.

Second, they did not approve the geothermal heating system! Why? Because of Fannie Mae guidelines, it would be difficult to resell the mortgage. What guidelines? "Well, it's about homes being different from the standard." This was so outrageous that I told them that they *had* to reconsider this. I wrote a letter stating my case. I had to rewrite it four times before I cooled off enough to write a respectful letter with the following points:

1) The EPA has stated that geothermal systems increase the value of a home $20 for every dollar saved in heating costs per year. If I had an oil burner, and required 1,000 gallons (3,785 liters) per year at $3/gallon, the increased value would be $60,000.

2) The home would be *easier* to sell because it has no yearly heating costs for fuel oil.

3) If fire is a concern, consider that with a GHP we would not be burning anything, therefore fire or carbon monoxide gases will be highly improbable.

4) It would be a sad thing if this bank went against our national need to reduce our dependence on foreign oil. I quoted the EPA: "If one million homes used geothermal heat pumps instead of oil, 21.5 million barrels of crude oil would not have to be imported each year."

They backed down. Construction started, was completed and we moved in. Was my pushiness worth it? Absolutely! We had no fuel bills. It was one of the smartest decisions I ever made. And the higher oil prices go, the smarter I seem to get.

All of my initial problems resulted from lack of knowledge. Many people I talk to have never heard of geothermal heating, or have incorrect information. It is an underutilized technology. Well over a million housing units are started each year in the United States, but only 120,000-plus geothermal heat pumps were shipped in 2008, according to the Department of Energy, and that includes not just homes but also schools and commercial buildings. How many of those new homeowners are even aware of what can be done? ✦

If one million homes used geothermal heat pumps instead of oil, 21.5 million barrels of crude oil would not have to be imported each year. Source: EPA

Germantown, New York

Earth Smart Home

Henry Hudson recorded his landing at what is now the Ro-Jan Creek in Germantown on his way up the Hudson River in 1609 to get fresh water. In 1709, the Queen of England convinced hundreds of Palantine Germans to settle here in order to produce pitch for wooden ships.

Germantown today is a small, interesting, rural town on the Hudson River of about 4,000 residents and 32 square miles (83.5 sq km). The Catskill Mountains, on the other side of the river, are very visible to the west.

This is where we found a partially built shell of a home on a former apple-tree farm. The building had just barely been started, but due to a divorce, was never finished. It was just left as an eyesore. When my wife, Martha, received her MFA degree at Bard College a few years before, she never dreamed that we would someday be living within a few miles of that campus.

Details: Single-story home, essentially new construction, 2,400 square feet (222 sq m), two-zone heating/cooling, 4-ton ECONAR GHP open-loop system with the main water well as heat source, and ductwork for hot and cold air delivery. The work was completed in 2007.

Open loop means that instead of having a water loop in a continuous cycle picking up heat and dropping it off, water from a source such as a well is piped into the GHP, releases its heat and is then piped back into the well again. The advantage is a much lower installation cost since the excavation for the pipes must be done for the water supply anyway. The disadvantage is a higher operating cost since the submersible well pump runs more often.

We found an architect in nearby Massachusetts and relayed our needs and wants. Needs: an attached art studio of 800–1,000 square feet (74–93 sq m), a ground-based geothermal heating/cooling system, and a single-floor, open interior with exterior views. Wants: a garage with a steel roof. When the final costs came in, some reductions had to be made: the garage was dismissed and other changes made, but the GHP and the studio were firmly in the plan. The bank resisted the GHP, but wilted when challenged.

Do we love it? Definitely yes! We have no fuel-oil bills. We have very comfortable heating and cooling plus partially free domestic hot water.

Chapter 3

How a Geothermal Heat Pump (GHP) Works: Borrowing Heat From the Earth

If you dig down about five feet or so in the ground to below the frost level, you will find the ground temperature to be amazingly constant, about 50°F (10°C) or higher (or lower), depending on the latitude.

It is cooler than the air in the summer and warmer in the winter. The earth's subsurface is an enormous heat sink—a solar collector—and it takes a large amount of energy to keep it in equilibrium. This heat energy comes in great part from the sun, a renewable and inexhaustible source of energy. In lesser amounts, it also comes from the center of the earth that we now know is a heat generator. The inner core of the earth is primarily made of a solid sphere of iron within a larger sphere of molten iron. Calculations show that the earth, originating from a molten state many billions of years ago, would have cooled and become completely solid without an energy input. It is now believed that the ultimate source of this energy is radioactive decay within the earth that continues to this day; the decay produces gradually diminishing temperatures from the earth's center to the surface.

This does not mean that dangerous radioactivity is a hazard to us. We can tap into all of this heat energy, transfer it into our home for heating and return that energy back to the earth during cooling: thus we are really *borrowing heat from the earth*.

Geothermal units use the same 100-year-old technology found in your refrigerator. They are both devices that move heat energy. It is worth noting that the refrigerator is the most reliable, longest-life appliance in your home. As Figure 3-2 and the inset spells out, a refrigerator removes heat energy from food and moves it into your kitchen. A geothermal system removes heat energy from the earth to heat your home and in the summer removes heat energy from inside your home back to the earth.

Heat naturally flows "downhill" from the warmest medium to the coolest medium. A heat pump is a machine that causes heat energy to flow in the direction opposite from its natural tendency, or "uphill" in terms of temperature. Because work must be done (energy must be applied) to accomplish this, the name heat "pump" is used to describe the device.

Average Annual Ground Temperatures in North America

City	Temp	City	Temp
Anchorage, AL	40°F / -4°C	Memphis, TN	63°F / 17°C
Austin, TX	71°F / 21°C	Mobile, AL	70°F / 21°C
Boston, MA	50°F / 10°C	Pittsburg, PA	52°F / 11°C
Chicago, IL	51°F / 11°C	Portland, ME	48°F / 9°C
Cincinnati, OH	57°F / 14°C	Portland, OR	52°F / 11°C
Concord, NH	50°F / 10°C	Reno, NV	50°F / 10°C
Denver, CO	51°F / 11°C	Witchita, KS	59°F / 15°C
Fargo, ND	42°F / 6°C		
Helena, MT	47°F / 8°C	Calgary, Canada	42°F / 6°C
Jacksonville, FL	64°F / 18°C	Halifax, Canada	45°F / 7°C
Little Rock, AR	52°F / 11°C	Vancouver, Canada	53°F / 12°C
Los Angeles, CA	64°F / 18°C		

Source: US Soil Conservation Service

Figure 3-1

A refrigerator and a heat pump are about the same physical size, are usually contained in a single enclosure, have similar components (compressor, evaporator, etc.) and both transfer heat energy. And they each require a refrigerant, which is a material used in a heat cycle that undergoes a phase change from a gas to a liquid, and back again.

But you will not find a gleaming white exterior nor an ice-maker on a heat pump. I view a geothermal heat pump as a refrigerator on steroids because it is designed to transfer a much higher level of heat energy. This requires a larger and more sophisticated compressor. In addition, a GHP is instantly reversible to provide heating or cooling. Of course, while a refrigerator is "plug and play," a GHP must be connected to the home's air ductwork or a hot water system, and to thermostats, water pipes, pressure gauges and valves. Oil and gas boilers also require electricity to run, but they need a constant supply of fuel in storage, a huge amount of oxygen to be burned, a chamber to hold the combustion, a heat exchanger to extract the heat energy and an exhaust system to get rid of the combustion products.

The most important difference, however, is in the efficiency. No refrigerator or oil burner is 100% efficient because like any machine, they have losses. Efficiency is basically output divided by input. Achieving greater than 100% efficiency means that you are getting out more energy than you are putting in. But surprisingly, a GHP can and does exceed that limit to provide 400% or even greater efficiency. Chapter 16 explains how this works in more detail.

How It Works: Heating

The major advantage of a geothermal heat pump is that it uses a small amount of electrical energy to transfer a much larger

How a Refrigerator Works

Like a refrigerator, a geothermal heat pump simply transfers heat from one place to another. When a refrigerator is operating, heat is carried away from the inside food-storage area to the outside area—your kitchen. Cold is not being added; heat is being taken out.

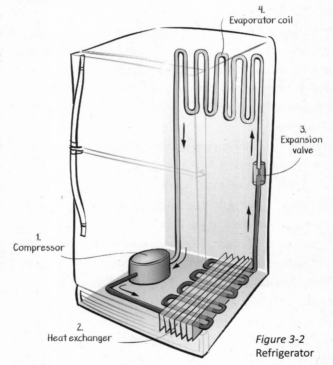

Figure 3-2
Refrigerator

To understand the operation of geothermal heat pump, it helps to understand how a refrigerator works. A refrigerator uses a refrigeration loop with four components:

1) A compressor
2) A heat exchanger
3) An expansion valve (often called a TXV, Thermostatic Expansion Valve)
4) An evaporator coil

A refrigerant is pumped through the loop to transfer heat energy from the inside to the outside.

The compressor (1) is a pump (basically a motorized piston) that pressurizes the refrigerant. Since temperature and pressure are directly related, as the pressure increases the temperature also increases in direct proportion.

The high temperature, high-pressure gas flows from the compressor to the heat exchanger (2) that is located outside the refrigerator. The heat energy in the heat exchanger is transferred to the cooler air in the kitchen and the refrigerant is condensed into a liquid.

Now the refrigerant is forced through a small orifice called thermostatic expansion valve or TXV (3). The small hole creates a pressure difference between the two sides of the device. This is like a dam on a river with a hole in the dam. Water leaking through the hole is at low pressure on the downstream side, but the water on the other side (being held back by the dam) is at high pressure. Again, the pressure/temperature relationship (low pressure/low temperature now) creates a cold, low-pressure liquid that flows through the evaporator coil. Relatively warm air from the food inside the refrigerator passes over the evaporator coil (4), giving up its heat energy to the cooler refrigerant. The refrigerant evaporates into a gas and the cycle starts all over again.

amount of ground-heat energy. In the process, most of that electrical energy is also converted to heat energy.

To make this happen, a geothermal system uses three loops to capture and transfer the earth's heat energy *(Figure 3-3)*. Each loop carries heat energy to the next loop. The **first loop** is a series of closed-loop pipes buried in the ground below the frost line in either a horizontal or vertical configuration, or placed in a lake or pond. Cool water from the unit in the house circulates through the ground pipes and the warmer earth releases heat energy into the water as it travels back to the GHP unit in the house. Non-toxic antifreeze is added to the ground loop water since it can sometimes be cooled below freezing by the GHP before the water circulates to the ground loop.

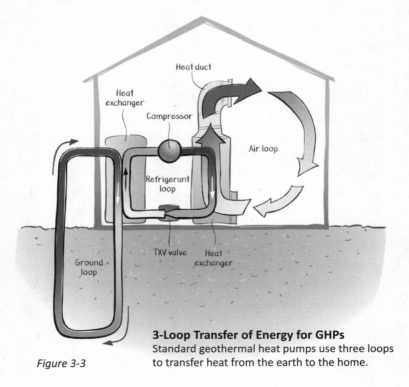

Figure 3-3

3-Loop Transfer of Energy for GHPs
Standard geothermal heat pumps use three loops to transfer heat from the earth to the home.

Or, as in my case, it is also possible to use an open-loop system by piping some of the well water directly through the GHP unit and returning it back to the well. There is no contamination of the well water; it merely circulates through plastic pipes before emptying into the well.

The warmed water from the earth passes through a heat exchanger (coaxial copper pipes) to the **second loop** containing an even cooler liquid refrigerant. Heat transfer takes place which cools the incoming water and sends it back to the ground loop to pick up more heat energy. The refrigerant accepts the heat energy and becomes a gas as it heats up. The now gaseous refrigerant is sucked into a compressor where it is compressed and superheated to about 165°F (74°C). This large and important jump is discussed in more detail in chapters 16 and 17.

In a water-to-air system, the heated refrigerant then passes through a radiator-like heat exchanger (air coil) over which air is passed. Or in a water-to-water system, 120°F (49°C) refrigerant is passed on to a hot water heating system by means of a heat exchanger (hydronic coil) that connects to baseboard registers or in-floor heating tubes. Air delivery ductwork of course, permits air conditioning. Again, Chapter 16 provides a more technical description of how the Gas Laws and the Three Laws of Thermodynamics come into play.

After the refrigerant transfers the heat to the air coil, it goes through a thermostatic expansion valve (TXV) and the pressure is released. The refrigerant becomes very cold (sometimes below freezing) as it circulates back to pick up more heat from the ground loop.

The **third loop**: In a water-to-air system, the fan-driven cool air is heated as it passes over the air-coil heat exchanger and is then ducted into the home at about 105°F (41°C), transferring its heat energy to the walls, atmosphere, you, etc. As the room air cools, it is returned to the GHP to pick up more heat energy.

How It Works: Cooling

Cooling is a similar process except that with just one tap of a button on the thermostat, a clever reversing valve sends the hot compressor output to the returning ground loop instead of to the indoor air loop.

While a geothermal heat pump operating in cooling mode uses essentially the same theory of operation as a residential AC system, there are two major differences.

First, a standard AC system requires a large outdoor heat exchanger in your backyard to dump heat energy into an already hot and, in some locales, humid atmosphere. This works, but it is very inefficient in that it takes a lot of electrical energy. The GHP needs no outdoor box; all components are underground

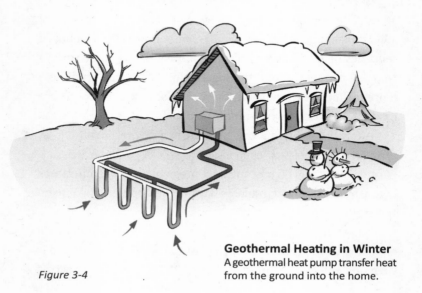

Figure 3-4

Geothermal Heating in Winter
A geothermal heat pump transfer heat from the ground into the home.

and inside the home. In my house, I send cold well water to flow over the heat exchanger pipes, heating the water which then flows back into the well, which in turn puts heat back into the ground.

Second, a standard AC system and a fuel burner/boiler are two separate systems connecting only at the ductwork. The GHP is a single system that can be reversed with a simple switch to provide cooling or heating. This reduces costs and increases efficiency. Alternately, we can set the thermostat for a given temperature and let the system determine if heating or cooling is needed.

That is it—a very simple concept, just three loops to bring in comfortable, quiet heat and cooling using well-established technology. The next chapter goes into more detail to describe different solutions to access that ground heat energy. ✪

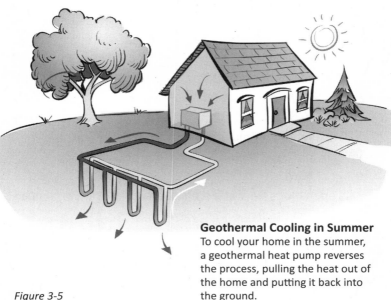

Figure 3-5

Geothermal Cooling in Summer
To cool your home in the summer, a geothermal heat pump reverses the process, pulling the heat out of the home and putting it back into the ground.

Chapter 4

Ground-Source Solutions: Open vs. Closed Loops

The choices between the various methods of tapping into the ground for heat energy will in most cases be determined by the size of the home; the size of the GHP unit; the amount of available land; the type of ground (clay, wet, ledge); well capacity; availability of a pond, lake or ocean; number of degree days and your budget. In fact, it is possible that there may not be many choices; there may be only one good solution. I was able to get an experienced geothermal contractor to sort this out. That is exactly what you need. Being knowledgeable about the terminology and your options will be to your advantage.

The solutions for moving energy from the ground into your home fall into two categories, open loop or closed loop.

- In **open loops**, the water is not recirculated; it is just pumped back to the source.

- In **closed loops**, the same water is constantly recirculated within sealed pipes/tubes.

The piping for either system will be capable of transferring heat from the geothermal heat pump into hot water radiators or radiant heating (water-to-water), or to the air duct system (water-to-air). Some homes even have a hybrid of both setups.

Open-Loop Systems

Open-loop systems (sometimes referred to as pump-and-dump systems) are installed less frequently, but they can be the least expensive method if ground water or a high-capacity well is available. They are the simplest to install and have been used for decades where local codes permit. This type of system uses ground water from an aquifer or pond which is piped directly to the building's pressure tank and then to the heat pump. The GHP manufacturer specifies the gallons per minute (gpm) or liters per minute (lpm) required. If you need a well capacity of 8 gpm (30 lpm) for a 4-ton GHP unit and 2 gpm (7.5 lpm) for domestic usage, then a 10-gpm (38 lpm) capacity well can provide both needs.

Once the water is cycled through the geothermal system, it is returned to the aquifer by discharging it through a properly sized drain field, a pond, river, lake, another well or the same well. A water filter is required to keep out contaminants and must be cleaned regularly. City and county regulations may be imposed for this type of system.

Water quality is a major consideration for open loop systems. If very low pH (< 6), very hard water (>100 ppm as $CaCO_2$), or hydrogen sulphide (rotten egg smell) characterizes the water, careful analysis should be done before using it in a heat pump.

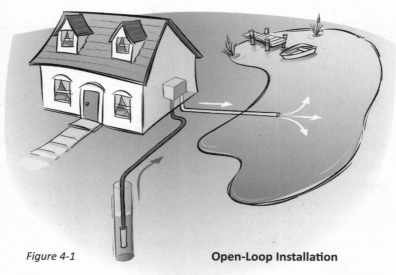

Figure 4-1   **Open-Loop Installation**

An open-loop system uses ground water from an aquifer (well) or pond which is piped directly to the building's pressure tank and then to the heat pump. Once the water cycles through the geothermal system, it is returned to the aquifer by discharging it through a properly sized drain field, a pond, river, lake, another well or the same well.

Why Are Open Loops More Efficient?

It turns out that open-loop systems always have higher efficiencies than closed loops. Why is this? The answer is that water conducts heat better than dirt. It's as simple as that.

Closed-loop systems depend on getting heat from sealed tubes that run through the earth. Open-loop systems send water into the geothermal heat pump so the heat can be extracted.

The length of pipe needed (called **loop length**) is a function of system size, climate, soil/rock thermal characteristics and loop type.

Closed-Loop Systems

In a closed-loop system, the fluid keeps circulating, picking up heat energy and transferring it, in an endless cycle as long as the pump is turned on.

Pond/lake closed loops are very economical if the water is fairly nearby. Lake loops are, for the most part, easy to install and are cheaper than horizontal ground loops because excavation is minimized and because water transfers heat energy more efficiently than a loop buried in the earth. Coils of plastic pipe *(see slinky loops)* are often bound together with ties, weighted down with cement blocks, filled with antifreeze solutions and simply placed on the bottom of the pond or lake. Generally a 6- to 8-foot (1.8–2.4 m) minimum depth is required. Typical minimum lengths are 300 feet (91 m) per ton of GHP capacity (a ton being equal to 12,000 Btu/hour).

Coils of plastic pipe ready to be buried in the lake. Photo courtesy of GeoSystems, LLC

Horizontal closed loops are considered if there is adequate land surface. Pipes are placed in trenches 6 to 8 feet (1.8–2.4 m) in depth in lengths that range from 100 to 400 feet (30.5–122 m), or more. The general rule is 500 feet (152 m) of pipe per ton of capacity, although this varies depending on many factors. A small circulating pump, about 1/6th hp (.12 kW), is also required.

Horizontal ground loops in the trench before burial. Photo courtesy of Peeples, LLC

Figure 4-2

Closed-Loop Horizontal Installation
If there is adequate land near the home, horizontal trenches can be dug 6 to 8 feet in depth and 100 to 400 feet in length to hold the ground-loop piping.

Winkler, Manitoba, Canada

Earth Smart Home

A modest Canadian bungalow of 2,080 square feet (193 sq m) including the finished basement, plus a two-car garage, sits on an exposed prairie near the town of Winkler, Manitoba. Their 5-ton, two-zone, horizontal closed-loop Northern Heat Pump system is a bit unusual in that it is multifunctional. First, in a water-to-air mode, the system delivers conventional forced air through ductwork for heating and cooling. At the same time, it has a water-to-water mode for radiant floor heating.

Radiant heat is a very comfortable approach because, for one thing, it is radiant instead of conductive. In addition, since heat rises, it is best to have the heat source as low as possible. The floor is as low as you can get. However, a radiant system is slow to heat and slow to cool because it is a slab of massive concrete. Now, it would be great if it were possible to insert chilled water into the floor in the summer to get room cooling via that floor—but, alas, it just won't work. When you cool air, it cannot hold as much moisture. The water vapor condenses out and becomes a problem. So a forced air system is needed for air conditioning. That also has condensation, but it is located at a single point so that it can be drained.

For Jacob Rogalsky, the issue is economy. He averages $85 (Canadian) a month in electric costs for heat, air conditioning and hot water all year. His neighbors with oil heat average twice that every month. "It's reliable. It beats the fuss about getting timely delivery of oil or propane when you live in the country. It's quiet. It doesn't take up much space. But the most important thing is that it saves so much money," Rogalsky says.

Photo courtesy of Northern Heat Pump

One of the newest approaches, **a slinky coil technique**, is growing in popularity since it requires less trenching. A slinky coil is flattened, overlapped plastic pipe in a circular coiled loop *(see photo on page 72)*. It concentrates the heat transfer surface into a smaller volume and thus reduces land area requirements by two-thirds. Large amounts of pipe are needed with this design. For example, one contractor provided 100 feet (30 m) of trench with 900 feet (274 m) of pipe per ton of capacity.

Figure 4-3

Closed-Loop Slinky Coil Installation
This installation method requires less trenching
but large amounts of ground-loop piping.

The choices between the various methods of tapping into the ground for heat energy will, in most cases, be determined by the size of the home; the size of the GHP unit; the amount of available land; the type of ground (clay, wet, ledge); well capacity; availability of a pond, lake or ocean; number of degree days and your budget.

Up to this point, I have shown that horizontal closed loops have required trenches in order to insert the piping. But there is another closed-loop approach that is usually less costly and certainly a lot less messy.

Where there is adequate space for a horizontal loop but a desire to minimize disruption on the surface, the **trenchless horizontal bore loop** may be the preferred solution. Using special equipment that bores holes horizontally under the surface, the operator directs the machine to drill at a slight angle down to a typical depth of 10–12 feet (3.0–3.6 m). Using the right technique, he/she can "steer" the drill head to go deeper or shallower, or turn right or left. The drill head emits a radio signal that can be detected by a special device that tells the operator exactly where and how deep it is.

A small-diameter tunnel is created underground by displacing soil with pressurized water. The operator drills the horizontal bore, then directs the drill bit to come back to the surface, typically, about 200 feet (61 m) away. The drill bit is then removed and two ends of the loop pipe are attached to the drilling pipe and it is pulled back through the bore hole, burying the piping underground.

Horizontal directional drilling in action.
Photo courtesy of A-One Geothermal, Ealham, Iowa.

Enfield, Connecticut

Earth Smart Home

This 2,100 square-foot (195 sq m) Cape Cod-style home in Connecticut had an inefficient, 33-year-old oil furnace that cost the homeowners $3,300 every year for heating and hot water. The owners, avid followers of alternative energy solutions, decided to go geothermal for both philosophical and economic reasons.

Terraclime Geothermal of Florence, Massachusetts, first conducted a detailed Load Calculation study using a "Manual J" software program. This is a common industry term that represents an analysis of the residence to determine the heating/cooling load. That, in turn, determines the size of the heat pump and then, finally, the configuration needed for the loop field. At one time this was a back-of-the-envelope calculation, but now experienced system contractors use a computer program to organize and compute the data. A contractor should be obligated to provide this analysis to the owner as a part of the bid. Terraclime not only did this, but they provided a comprehensive Energy Cost Analysis which compared their oil burner costs to the new heat pump installation costs.

The total installation, completed in 2008, took only five weeks. Their 3-ton DX system was made by Advanced Geothermal (ECR Industries) and six 70-foot (21-m) boreholes were used for the ground loop. The result: the GHP provided heat throughout that first winter without relying on the backup oil burner. Annual costs were reduced to $583 (for electricity to operate the GHP) and that included winter heat, full air conditioning, dehumidification, and partial hot water. In 2008 their Federal tax credit was limited to $2,000 (now increased to 30% with no limit) plus the State of Connecticut provided a $1,500 rebate.

The owners are pleased to report minimal maintenance, "All I have to do is change the air filters every six months—much less work than writing a check to the oil company each month."

This technique of horizontal directional drilling (HDD) allows the loop to be placed underneath homes, basements, driveways, wooded lots or even swimming pools. The only digging required is for the header unit and the supply/return piping into the house. This type of loop is especially attractive in a retrofit situation which minimizes disruption to the landscape. Please note, however, that clay soil may not be appropriate due to air gaps in the soil. And care must be taken when pulling piping through the hole since hitting sharp rocks can result in system leaks.

Not all contractors have this boring equipment. To locate a contractor in your area that uses this technique, do an online search for "geothermal horizontal drilling." Make certain that the driller has completed many such installations.

Vertical-closed loops are often needed if land surface is limited. Drilling equipment bores vertical small diameter holes 100 to 600 feet (30–183 m) in depth to hold the plastic piping. This is usually the most expensive approach because of the drilling expense. However, it is very useful when the land is rocky. The number of holes and depth varies depending on many factors, including soil and rock condition. Some locations require 180 feet (55 m) of bore hole per ton and 360 feet (110 m) of piping. The closed loops contain water or a mixture of water and antifreeze. A special grout is often pumped down the holes to surround the pipes and insure good heat transfer from the surrounding earth.

The **Direct Exchange (DX)** approach to ground-based geothermal heating/cooling is unique enough to merit a separate chapter, which follows. ✪

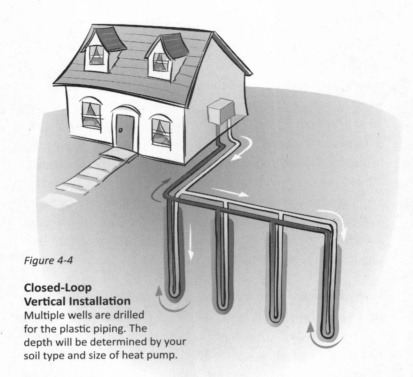

Figure 4-4

Closed-Loop Vertical Installation
Multiple wells are drilled for the plastic piping. The depth will be determined by your soil type and size of heat pump.

West-Central Minnesota

Earth Smart Home

I remember Minnesota. It is a great place with good people. I also remember having my car tires freeze flat on the bottom, causing a terrible thump/thump/thump. It does get cold there. If a GHP system works there, it will work anywhere.

To prove the point, I want to describe a West-Central Minnesota residence that proves that a GHP system installation can be flexible enough to adapt to local topography and individual architecture.

Built in 2004 for a family of six, this home has 4,300 square feet (399 sq m) of space on two floors plus a basement and garage adding another 864 square feet (80 sq m). They installed three GHPs, with a total 10 tons of capacity. The manufacturer was ECONAR, a Minnesota manufacturer and the same company that supplied my unit.

The ground source was planned to be 150- to 165-foot (46–50 m) deep boreholes but they hit bedrock at 135 feet (41 m). It was better to dig more holes than dig deeper, so they ended up with 13 boreholes plus a well for water.

They have three zones: 1) Top floor: water-to-air to provide heat and AC; 2) Main floor: water-to-air to provide heat and AC; 3) Basement and garage: water-to-water for radiant heat.

A unique feature is that they installed one of the GHPs on the top floor since it is more efficient to create the cool air closer to where it is needed.

The total cost of $39,000 included the boreholes, ductwork, a backup gas furnace (required by the local utility co-op in order to get off-peak electric rates), radiant heat in the basement and garage floor, a desuperheater, a backup water heater, and a Heat Recovery Ventilator (this permits a tighter home to save additional costs while still circulating fresh air in every room).

If they had used a standard propane heating system, it would've cost approximately $5,400 per year (at $2/gal for LP). The average annual electrical cost just to run the GHP units is about $1,200. So they are saving close to $4,200 every single year. At the same time, their home is very comfortable in every season and with the desuperheater, they have free hot water in summer (and almost free in the winter) for six people. They are especially happy with the good, consistent humidity and the even distribution of heat and cooling. This was a very successful installation that is the result of selecting a highly qualified installer with over 15 years of experience.

Chapter 5

Direct Exchange (DX) Geothermal Systems

If you are considering a geothermal heat pump, then you would be smart to become familiar with the DX, Direct eXchange concept (sometimes referred to as DGX, Direct Ground eXchange). This closed-loop system simplifies the standard three-loop approach by eliminating one loop, while still providing heating, cooling and even full domestic hot water. While it may not be the least expensive in terms of installation, comparative testing has shown a 20 to 30% increase in efficiency with DX systems.

In the preceding chapters, I have described the three loops used to transfer heat energy from the ground into your home. The DX system combines two of those loops by using a refrigerant in the ground loop. Annealed copper tubing is used instead of plastic and it is grouted and sealed with a heat-conductive compound to protect it. This elimination of a loop and its related equipment has immediate advantages:

- Increased system efficiency due to one less heat-transfer loop and the use of high-pressure refrigerants.
- It takes less energy to operate, but requires more refrigerant (to fill the underground copper pipes).

tion pump and the energy to run it since the compressor does all the work.
- It reduces excavation and installation costs thanks to the higher conductivity of copper pipe. This means less pipe is required (smaller diameter and shorter in length), and fewer, shorter boreholes will be required.

A discussion of efficiency and how it is measured is presented in Chapter 16. The US EPA Energy Star Heat Pump criteria set a minimum efficiency for DX systems at 350%. Websites of DX manufacturers show performances ranging from 400% up to 580%, depending on the configuration. In addition, many DX systems will provide full domestic hot water. The downside of

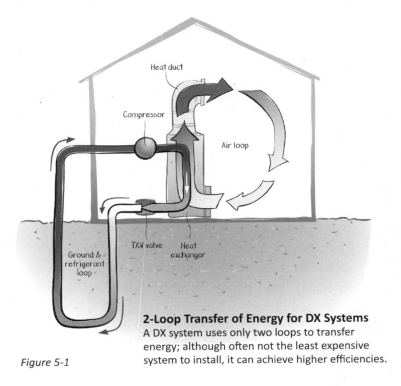

Figure 5-1

2-Loop Transfer of Energy for DX Systems
A DX system uses only two loops to transfer energy; although often not the least expensive system to install, it can achieve higher efficiencies.

DX systems is that qualified designers and experienced installers are harder to find.

When DX systems were first introduced in the 1970s, there were some failures due to poor installation procedures. In those early DX systems with vertical boreholes and very long, high-pressure refrigerant loops, some of the oil from the compressor moved into the refrigerant, thus reducing the lubricating capability of the compressor and damaging it. In addition, copper pipe fractures in the ground loop caused by ground heaves (due to the earth being frozen) allowed refrigerant to escape into the soil and ground water, posing a threat to the environment. These were serious and expensive problems for the homeowners.

It is important to note, however, that the technology and installation techniques have greatly improved. While not the most popular, large numbers of DX systems are in operation in the United States and Canada. For example, as of 2010, one company in New York—Total Green Geothermal *(www.totalgreenus.com)*—has over a hundred systems in place, all operating without problems.

DX System Attributes

- Standard vertical and horizontal loop fields are seldom used in DX systems. Instead, a less invasive approach utilizes 30-degree angled shafts, each less than 3 or 4 inches in diameter *(see Figure 5-2)*.
- Special brazed copper is now used and tubes are adequately spaced apart. Pressure testing is a must.
- The empty space around the copper pipe is filled with heat conductive, cement-like protective grout.
- Oil separators or other solutions are a part of all DX systems to insure proper oil return to the compressor.

- Although ground acidity is not common, most DX companies use a procedure that measures the pH of the soil 3 feet (0.9 m) below the surface where the manifolds are to be located. If the pH is 6 or less, the loops may get cathodic corrosion protection (much like a boat propeller in seawater). Installing DX systems in limestone may be a bad idea.
- Although special techniques are required to drill in wet areas, the results can be worth it due to better heat transfer.
- Professional installation for any geothermal system is critical, but even more so in a DX installation. Leakage, cavitation and pump failures can still happen from improper installation. If a leak should occur on the DX earth-loop side, the repairs and refrigerant recharging would be a substantial expense. Check the warranty on the installation, not just on the unit itself. Work only with a reputable contractor/installer with lots of experience installing DX systems. ✪

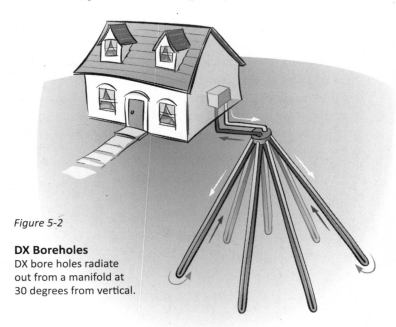

Figure 5-2

DX Boreholes
DX bore holes radiate out from a manifold at 30 degrees from vertical.

Cuddebackville, New York

Earth Smart Home

This stone house in Cuddebackville, New York, is unique for several reasons. Built in 1753 when King George of England ruled the colony, it was converted to a home in the 1930s with an oil-fired steam radiator. Now it could certainly be the oldest home in the US retrofitted with a DX geothermal system.

Because of its age and construction, special consideration had to be given to the building before any heating installation was started. After an evaluation by a Certified Building Professional, Total Green Geothermal of Monroe, New York, did a GHP software study to determine the heating requirements, how best to distribute the heat, and how to create a "thermal envelope" that included insulation of below-grade walls in the basement, high-density fiberglass in the attic, and air sealing the living space. Normally, a GHP delivers quiet, low velocity warm air to the living spaces. But, in this case, to retain the unique character of the building, a special smaller, high-velocity ductwork was chosen.

Heating bills prior to the installation were over $5,000 per year. Now it costs $65/month for heat, hot water and full air conditioning. For a musician, this is especially important in order to protect sensitive musical instruments. The owner says, "I am going to recoup my investment in seven years. Next, I plan to install solar panels." The 5-ton DX unit was built by Advanced Geothermal Technologies in Reading, Pennsylvania. Ten holes, each 70 feet (21 m) in depth, were drilled at a 30-degree angle.

It is good to see an older building in such excellent use today.

Photo courtesy of Total Green Geothermal, LLC

Chapter 6

Using Your Well as the Heat Source

I was intrigued the day my geothermal contractor suggested using our well to transfer heat to our proposed geothermal heating system for our-yet-to-be built home. I had mentioned that I had just learned that our new well had a capacity of 12 gallons per minute (45 lpm). It seemed to be the perfect solution.

After almost four years of use, I consider it a very good approach for our situation. But like most things, every approach has its tradeoffs. The major advantage of an open-loop well-source system: lowest installation cost. Major disadvantage: higher electrical cost to run.

I recently read an article about a homeowner who had installed a well-based geothermal system in an existing home. Naturally, I could not wait to read it in order to compare ideas. He was very enthusiastic about his system but had real problems in getting it to work. An important lesson can be taken from that article. First, he clearly chose a contractor with no experience in the installation of a well-based system. That cost him discomfort, time and money. One thing the contractor missed was the fact that in conditions of very prolonged cold weather, the heat taken out of the well can, at times, cool down the water and make the system less efficient.

When we use ground trenches or deep holes for heat, we are tapping into an almost infinite heat sink. But an aquifer is smaller and more subject to changes. My well-water temperature runs about 50°F (10°C) in the summer as I add heat that was extracted from the cooling of our home, and about 43°F (6°C) in the winter as I transfer heat energy out of the well.

My geothermal contractor and my well contractor both recognized that cooling of the water in winter was a potential problem and planned a solution that was implemented from the start. They installed a third pipe from the GHP, through the basement, past the well and toward a stream in the rear of the house. When the well water temperature reaches 40°F (4°C), the system automatically diverts the really cold water to the third pipe. It only runs in that mode for an hour or so and then switches back. It is so quiet that I never know it is happening unless I am passing near the outlet. It is double piped for weather protection and has a grid at the end to keep out wildlife.

There are some who are concerned about the impact of a well-based system on the aquifer. Potable underground water is a precious resource. The use of plastic pipes in place of copper is commonplace now for very good reasons. But there is concern that they may be adding pollutants to the water, especially if large numbers of homes are involved. My research shows that this concern is probably unwarranted.

....

According to data supplied by the US Department of Energy Office of Geothermal Technologies, nearly 40% of all U.S. emissions of carbon dioxide are the result of traditional heating, cooling, and hot water systems in residential and commercial buildings. This is roughly equivalent to the amount of carbon dioxide contributed by automobiles and public transportation.
....

The materials used are proven safe: cPVC (chlorinated polyvinyl), HPDE (high-density polyethylene) and PEX-a and PEX-b (which are "cross linked" polyethylenes) are commonly used water piping. Each have been approved and certified for use in drinking water applications by NSF International and ANSI, the American National Standards Institute. However, scientists at the University of Virginia have reported to the American Chemical Society that their studies have shown that some plastic pipes (in particular the HDPE pipe) can affect the odor and/or taste of drinking water. cPVC and PEX-a have a low odor potential and did not release organic chemicals. HPDE had the highest odor production, but did not release organic materials. PEX-b had a moderate release of odors and materials. All of this seems to disappear in a month or two of use. There is no data that shows any adverse health effects because of these materials.

I had a few more thoughts on this subject as I made my temperature rounds on a cold winter morning. It was our coldest outside air temperature that winter at 1°F (-17°C) and a wind chill at -8°F (-22°C). Inside it was a comfortable 70°F (21°C). We like to keep the night setback at 68°F and set the system to automatically move to 70°F at 6:00 a.m. I measured the timing at the changeover—it took less than five minutes. This told me that the system was working well—in spite of the cold. Then, in washing up, I was again struck by how really cold the water from the bathroom faucet feels at this time of year—43 degrees actually.

Well Source Advantages/Disadvantages

Advantages: Using a well as your source is considerably less costly than the trench or deep-hole approach. It is also lower than a traditional furnace and full-home air conditioning. This means that an immediate positive payback is gained because of the savings in oil costs and construction costs. Other aspects:

- Requires minimal maintenance; just clean the water filter 6 times a year.
- The process does not pollute the water.
- A well-based system uses more electricity in the winter months because the submersible well pump is running quite often. For my system, this amounts to about $100–$150 per month for four months. Since we only use summer air conditioning in very hot, humid weather, we rarely have additional electrical costs during those times. The net gain from the elimination of oil costs and oil furnace maintenance is far above this.
- The well temperature can be lowered in periods of very cold weather, requiring an alternate disposal pipe a few times each winter.
- Very dirty or contaminated water should not be used.
- Local codes/restrictions may apply. ✲

Homeowner Insurance Savings? I had this naive idea that homeowner's insurance companies should provide a discount or credit for GHP systems, based on the lower risk of fire or carbon monoxide emissions. However, when I checked with my insurance company and several underwriters, the answer was fairly uniform. First, they were not certain what a geothermal system was. And second, they stated that insurance companies do not give discounts for heating systems.

Chapter 7

Geothermal Advantages / Disadvantages

Of the dozen advantages listed on the next page, the driving force behind my decision was the first one—the urgent desire to eliminate fuel-oil costs. The rest are very real and very much appreciated, but they really come with the package.

We cannot ignore, however, the disadvantages listed below. How important is a higher installation cost? If a GHP for your home costs more than a standard heating and cooling system and you plan to move in the next three to five years, why bother with a GHP? Well, since home heating/cooling can be an expensive proposition, you would still profit from years of zero fuel bills. And if the extra cost difference is added to your mortgage, the added annual mortgage costs will be less than the annual cost savings. The payback is immediate. Then factor in the 30% federal tax incentive, and a possible state or province incentive. In addition, the increase in your home's value could more than pay for the difference.

GHP Advantages

- Fuel costs (oil, natural gas, propane) are zero, year after year.
- GHP systems conserve natural resources by using the earth's renewable energy.
- GHP systems provide heating *and* cooling and can change modes with a flick of the thermostat.
- GHP systems have lower maintenance costs, no oil filters or nozzles to clean, no open flames, no carbon monoxide, no chimney, no air pollution, no oil storage tanks, and no sludge.
- GHP systems are by far the most efficient (400–500%) system available in converting a small amount of electrical energy into a large amount of home heating/cooling.
- There is a notably quiet operation.
- When the system is running, a desuperheater water heater heats domestic water, saving additional costs.
- Long system life: you can expect 20–25 years of service, which is double that of a boiler or furnace and AC system. The plastic loop pipe is guaranteed for 50 years.

Slinky coils for a GHP system prior to burial.
Photo courtesy of GeoSystems, LLC

- Increased home value: See *Chapter 8 – Does a GHP Really Increase the Market Value of a Home?*
- GHP systems are more affordable now thanks to federal subsidies, tax credits and rebates which are available in the United States and Canada. See chapter 13 for details.
- There is a measure of satisfaction in watching the fuel delivery trucks pass by without stopping.
- You are granted bragging rights (or at least a dinner party conversational topic) about how you heat your home without fossil fuels.

GHP Disadvantages

- Most closed-loop geothermal installations cost more than a fuel burner plus full AC, perhaps 20–30% more. This is because of the drilling or earth excavation and the installation of the pipe for the vertical or horizontal ground loops. However, an open-loop heat source can often cost less than a fossil-fuel system plus full AC.
- While a GHP is simple in concept, compared to traditional heating and cooling systems, it is more complex to install. It has advanced electronics, sensors, control loops, and safety controls, which means that the installation and service personnel require a higher level of training. It's like the difference between maintaining your 1955 Ford by yourself as opposed the professional skill required to maintain a 2010 SUV. See chapter 15 for more information on this topic.
- While you are saving on fuel costs, you may experience an increase in electricity usage, depending on your configuration. Remember, though, a small amount of electrical energy adds a large amount of heat energy. ✪

Chapter 8

Does a Geothermal System Really Increase a Home's Market Value?

My original premise that a ground-based GHP will provide an increased market price was simple. It seemed intuitively logical that if two homes were equal except that one had a GHP system, the GHP home would be more desirable and a buyer would pay more. Then I found that the EPA had written the same thing more explicitly. They stated that the value increases by $20 for every dollar that is saved per year in heating costs. So if you would have used, say, 1,000 gallons (3,785 liters) of heating oil at $2.50 per gallon, a GHP system would save $2,500 dollars per year and that would increase the home's sale price by $50,000. This is not unreasonable. If a buyer added the increased cost as a part of a 30-year mortgage, the annual increase in payments would be less than the annual savings in fuel costs.

But the EPA does not explain how they came to this conclusion. And you could argue that the EPA does not establish market values—the marketplace does this. So I have wondered how realistic is this? How could anyone verify, or quantify this? Well, I could sell my home. That would give me one data point. But I'm not that crazy. Could I spend a scarce $375 to have a certified appraiser establish my market value? I could and I did.

Let me review how appraisers work. Ideally, they examine the database records to find three homes in your area that have sold over the past year, and are exactly like yours. Then they average the prices and presto—they have a market value. Of course, they will not find even one home just like yours or mine. In fact there are no homes with geothermal heating systems that have been sold in our area. So the appraiser finds at least three homes that are generally in the assumed price range, and sets up a series of increases and decreases in value to bring the homes into an assumed and approximate equality. Then that average would be a basis for the value of the appraised home. For example, if the comparable home had a two-car garage and mine did not, the appraiser could *lower* the value of the comp home by the value of a two-car garage. If I had a geothermal system and the comp house did not, the appraiser could *add* a value to the comp home. The question is—how much?

Here is what happened. Arthur was blindly chosen from a list provided by a mortgage company as a reliable, certified appraiser. He arrived and quickly started to sketch a floor plan using a laser rangefinder to get the dimensions so that he could add up the square footage. He then took interior and exterior photos with his digital camera. I gave him a basement tour showing the geothermal system quietly humming along (no chimney, no oil tanks). I showed him the water heater with its domestic cold water pipes "in" and hot water "out" on top of the unit and the geothermal hot water "in" and cooler water "out" at the base. I explained how I was saving at least 1,000 gallons (3,785 liters) of oil per season, and quoted him the EPA statement about the increase in value because of a GHP. He listened carefully to my pitch, and then said that there would be some people who would not want a GHP because it was too different—just what I did not want to hear. He did state that FHA guidelines permitted an increase due to energy-efficient systems. After he drove off, I was left with much uncertainty about how this would turn out.

It took a week before I received the details in writing. Before I show the results, I must explain two things. First, during the negotiations for a construction loan, my bank required a certified appraisal based on the actual price I paid for the land plus the architectural drawings for the home. This gave me a baseline that was three years old. Next, during the past year of occupancy, the value of US homes declined. In our county, it was small, perhaps 5%, but there was a decline in value.

The new appraisal showed a 14% increase in value! (Although I would not want the town assessor to read this.) That is about a 4.5% increase per year compounded. And in reference to the EPA estimate, my value increased by $25 for every dollar I saved in oil costs, even after I reduced those savings by the increased cost of electricity to run the GHP.

Appraisers are not buyers. It is impossible to know how important it is to a buyer to be free of fuel costs or whether the buyer understands that a GHP system is not "strange"—that it is basically established refrigerator technology.

What have I learned from all of this? I believe that a GHP system does increase the home value. How much of an increase will be unknown until there is an actual sale. In the meantime, we who have such systems have a good thing going for us. We are not about to sell. That tells us all something. ✪

Chapter 9

Learning to Live With a GHP System

Maintenance

You have a new GHP installation in your basement; a winter northeastern storm is coming on. The system is so quiet that you have to keep checking to see that it is working. You are a bit uncertain and nervous but you finally relax. Suddenly, the system quits! There is no heat. Your family looks at you. *Do something!* This was supposed to be so reliable! Doubts creep in. Have you made a huge mistake? Maybe a conventional furnace would've been better after all. Don't panic—call your contractor.

This happened to me three times. In every single case, the GHP system did exactly what it was supposed to do—it shut itself down because of preventable maintenance problems.

The first time was because of a dirty air filter. When we were in the drywall construction phase, my builder did not want to leave an unattended propane heater on all night for safety reasons. But it was freezing weather and sanding and plastering were underway. The GHP had been installed, but the well water was not yet hooked up. So I decided to use the electrical backup capability to provide heat. This worked fine, but even though the return ducts were disconnected, a great deal of dust got into the system, clogging the new air filter. Lesson learned: never turn on a GHP during construction. Keep the air filter clean on a schedule and record it on the maintenance sheet.

The second shutdown occurred because a mouse chewed through the well pump control-wiring insulation. That automatically turned off the GHP. Mousetraps are now in place and checked regularly.

The third time it was the water filter. Any new well often has particulate matter. It can easily be cleaned, as I found out. Set up a schedule and record it.

From that point on, the system has worked flawlessly, mousetraps included. By the way, I discovered that better mousetraps *have* been built.

There is one other lesson to be learned. If you need to clean the filters while the system is on, you obviously must turn the system off first. When you finish cleaning and turn it back on, you may not realize that there is a time delay of about five minutes to protect the compressor. Even after that it may not start. An age-old trick works every time—turn the GHP circuit breaker off and then back on. Get to know your circuit breaker box.

Backup Systems

In colder climates, GHP systems always include a built-in electrical backup heater that can take over in case of, say, an unlikely compressor failure or even as added energy in case of very, very cold weather. But the contingency that concerned me was a power failure in the middle of a serious snowstorm. Propane-fired emergency power generators are a neat answer but expensive for me. I picked a wood burning stove as my backup. So, I may be burning something after all!

Night Setbacks

With an oil burner, we assume that we can save energy costs by turning down the thermostat three to five degrees at night, depending on our comfort level. But with a GHP, I noticed that the second-stage compressor *(see next chapter for more info on stages)* would often kick in, adding costs and creating doubts about a setback. So I checked with my manufacturer who consulted with the thermostat manufacturer. The answer came back: "The more efficient the system, the less savings with night setbacks. The most efficient approach with a geothermal system is to set the thermostat and never change it." But I preset it back at night for comfort, not savings. Of course, if the home is going to be unoccupied for long periods, use a setback for savings.

Comfort Level

In our former home, the schoolhouse, the oil furnaces were in the basement just under our bedroom. We actually had three smaller furnaces—for the studio, first floor and second floor—instead of one large one. Every time that thermostat kicked on a furnace, we heard a loud roar as a sudden rush of air and combustion took place. We seemed to have problems with incomplete combustion because the nozzles often became clogged with residue from partially burned oil. And there was a transformer located near the combustion chamber that kept burning out. We needed an annual cleanup to maintain efficiency (it was certified as 86%, but that soon deteriorated). We needed 3,100 gallons of fuel oil each year even though we set the second floor temperatures low.

In contrast, our GHP is quiet. At times I can just hear the compressor running when it goes into stage two if I am in our dining area directly above the GHP unit. In addition to the quiet operation, the rooms are always comfortably warm in the winter. We do not use AC too often, but being close to the Hudson River, it does get warm and humid. When we need AC, it feels so good! There is also comfort in not having to worry about an on-site supply of fuel oil getting low or running empty. And finally, the absence of flames and pollutants is comforting because it reduces the odds of something bad happening.

Thermostats

I am somewhat nostalgic for the famous Honeywell Round thermostat. I think every home we've ever owned had at least one, including the schoolhouse, which had three. The bimetallic scroll worked without any integrated circuit; it gave us the current temperature, allowed us to set the temperature and turn it off or on. It was simple, low cost and doomed. That's because the on/off part was accomplished with a mercury switch—no longer a smart idea.

GHP thermostats are all digital. They hook into a digital circuit board and have more functions, though if you look at the outside case of a typical model, it seems simple enough.

You see just two buttons: they set the desired Set temperature Up or Down. On the left side of the display you see the Time which alternately flashes with the current Inside Temperature. In the middle is an AM/PM indicator. On the right is the Set Temperature. Below that is a Flame Symbol that indicates it is in the heating mode. And above the time is the Day of the week (WE for Wednesday). Not shown since the

Digital Thermostat

system is Off in the example: in the upper right corner would be an STG-1 or STG1-2 to indicate if only the first stage or both compressor stages are on.

Now, open the case. Six more buttons are exposed. It may look complicated, but the homeowner seldom needs to use all of these settings since most are only for the installer. This is a good reason why it would be smart to save your installation manual. Some directions are written on the inside of the cover, but you really need to follow your own specific thermostat directions.

Your thermostat may not be exactly the same, but it will have similar features. Here are the buttons for this particular thermostat:

① Raises temperature setting

② Lowers temperature setting

③ Time button (changes thermostat Time, up or down)

④ Program button (sets thermostat from factory presets to your desired settings)

⑤ Run (run program) button

⑥ Hold temperature button

⑦ Fan switch (ON, AUTO)

⑧ System button (Cool, Auto, Heat, Emerg, Off)

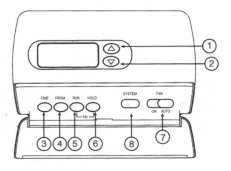

Figure 9-1   **Digital Thermostat Schematic**

When first turning on the system, you will want to correct the clock setting using the Time button ③ and learn how to correct for Daylight Savings.

Then you will need the System button ⑧ to set Cool or Heat, Emerg or Off. (Emerg= Emergency is used if the compressor will not work and you want to use the backup electrical strip.) We use the system button for comfort, knowing full well that it is more efficient not to adjust the temperature often.

Next, you may want to program the system for automatic temperature settings for day or night (refer to your manual). Here is where you set a heating/cooling schedule plan so that Weekdays, Saturdays and Sundays have a different schedules (or the same) and settings can vary throughout the day. You can always override the Program ④.

Still hidden behind the upper case are two AA cells for Time backup during power failure. The GHP will not work in a power failure but the clock will. ✸

A typical geothermal heating and cooling system that also provides free domestic hot water. Photo courtesy of GeoSystems, LLC

Chapter 10

Reducing Operating (Electrical) Costs

Geothermal systems require less energy than standard heating and AC because they are more efficient. But they still require electrical energy to work. The whole concept is to use a small amount of electrical energy to control a large amount of heat energy. So anything we can do to reduce electric power will increase system efficiency even more and cut monthly bills. The colder the weather, the more important this becomes.

Constant Pressure Controller

Since I am using the well as a heat source, my submersible pump replaces the electrical power that would be used to pump water through a ground loop. Our submersible pump is larger with multiple needs, and so it requires more power. The well contractor suggested a special control system that provides just enough power to meet the demand (instead of having the pump either completely on or off), while at the same time, keeping a constant water pressure. It gives just enough power to handle the current load. If I only use the shower, it provides enough power to provide water for that. If I have the GHP heat, clothes washer and dishwasher all running at the same time, it will increase the power accordingly. It works in conjunction with a small pressure tank and a microprocessor to send the right amount of electricity to the pump to speed it up when more water is needed, or to slow it down when less is needed, while keeping the pressure at a constant 50 psi.

Multiple-Stage Compressor

My GHP, like most units, has two levels, or stages, of compressor operation. A few systems have three stages. The GHP automatically decides when to change stages, based in part on the difference between the current temperature and the set temperature. Stage One is normal operation. Stage Two (which requires more power) goes into effect if the weather is particularly cold for a long period, for example. My goal is to keep the system in Stage One as much as possible. I am careful to only make small changes in settings at any one time in either heating or cooling mode. Unlike a standard compressor which only operates at full capacity, two-speed compressors allow heat pumps to operate close to the heating or cooling capacity needed at any particular moment. This saves large amounts of electrical energy and reduces compressor wear.

Scroll Compressor

Another advance in heat pump technology is the scroll compressor that replaces the piston design. It consists of two spiral-shaped scrolls. One remains stationary, while the other orbits around it, compressing the refrigerant by forcing it into increasingly smaller areas. Compared to the typical piston compressor, scroll compressors have a longer operating life, are quieter and provide 10–15 degrees higher temperatures. A scroll compressor has a digital controller and fewer moving parts (only 6, compared to a piston type with 20 parts). It is more efficient, thus saving electricity.

Heating Domestic Water with a Desuperheater

Most geothermal units have an optional feature called a desuperheater, which heats domestic water for showers, etc. This component consists of a refrigerant-to-water heat exchanger installed at the discharge of the compressor. The hot gas at this point is in a "superheated" condition. In the desuperheater, the refrigerant releases some of the heat into the cooler water through the copper walls of the desuperheater heat exchanger. A small pump circulates the water from the electric water heater/storage tank to the heat exchanger, where it picks up heat energy and returns it back to the storage tank.

This excess heat energy is available in both the heating and cooling modes. However, there is a greater benefit in cooling mode since there is more heat to be eliminated. The amount of hot water to be generated is a function of the run time of the GHP unit. On very hot or very cold days the desuperheater may be able to generate the majority of the domestic hot water needs because of the longer run time. A safety cutoff switch shuts off the circulator if the temperature exceeds 130°F (54°C).

The desuperheater can reduce your water heating bill by about 20–40%. By design, it provides only partial water heating and you will still need a water heater. In the dead of winter when both the desuperheater and the heat pump are working, the amount of energy given to the hot water tank is fairly small.

Desuperheater capacity is directly related to GHP capacity, so the hot water provided by a 5-ton unit would greater than a 3-ton unit. A word of advice: watch for scale buildup in the desuperheater exchanger that could affect its performance.

Other Water Heating Options

If you have an electric or gas water heater you know that it is energy hungry. It is expensive to keep the tank at 120°F (49°C) even when no water is drawn. I chose a desuperheater to reduce these costs; it uses spare heat from the heat pump to heat my water. But you should be aware of other solutions.

Solar Heating: Solar water heaters can be a great bargain for heating domestic water if your home has good south-facing exposure. They are not too common yet because of installation costs ($4,000–$8,000), but federal tax credits are available. They require a storage tank (various sizes are available), but with longer winter nights and cloudy days, a gas or electric auxiliary booster is needed. More details about the various system options can be found in the book, *Got Sun? Go Solar.*

Tankless Heating: On-demand tankless (or instantaneous) water heaters produce hot water only as needed, eliminating storage tanks with their standby losses and operating costs. Hot water never runs out unless the demand is greater than the designed flow rate (2–9 gpm; 7.5–34 lpm). These are independent of the home heating system and produce a constant supply of hot water on demand by using "flash" heating elements to reducing operating costs by 10–15%. They are the primary

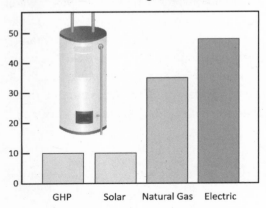

Solar and GHP heating of domestic hot water costs much less than natural gas and electric.

Figure 10-1

Source: VDI (Association of German Engineers)

source and need no auxiliary backup. Costs are $800–$2,000 (plus installation).

Air Source Heat Pump: Another approach is to use a small dedicated, stand-alone air source heat pump, with its own compressor that uses the air in the basement as a heat source solely for domestic hot water. A storage tank is required. They cost around $1,500–$2,000 (plus installation).

On-Demand (OD) Water Heating: A few companies have integrated "on demand" heating into their geothermal heat pump systems. HydroHeat *(www.hydroheat.com)* offers a unique patented On-Demand approach that allows you to generate hot water and to heat/cool your house with a single geothermal heat pump system, even when the heat pump is not at that moment running or when heating/cooling is turned off. This OD subsystem package within the heat pump takes advantage of the GHP compressor and uses a "Hot Water Priority" protocol,

---

Swimming Pool Heating

I assumed originally that this might be a natural additional use for a GHP installed in a home since in the summer there is excess heat being transferred out of the home and into the ground. Why not send it to the pool instead? There are such installations that work very well. But the AC is not always turned on in the summer and it turns out that a separate heat pump is often needed for the pool. If you are interested, look for companies who provide dedicated heat-pump pool heaters/coolers. Federal tax credits do *not* apply.

One of the best pool heaters I ever saw was in the French countryside. It used black plastic pipes to trap the heat from sunlight and a small circulating pump. If you have ever tried to use your garden hose after it has been sitting in the sun, you'll understand.

a second reversing valve, a double-walled heat exchanger and a water tank to provide hot water at all times. It is the primary heat source and no auxiliary electrical heater or gas burners are needed.

While an OD system and a desuperheater are both integral to a GHP, it is interesting to see the differences between them:

A **desuperheater** is "Home Heat Priority," that is, it only provides *excess* heat energy for domestic hot water. It does this only when the unit is running and providing heating/cooling. It is an auxiliary system and requires a primary, separate water tank heated by electricity or gas. It provides 20–40% of the hot water. The added cost is about $500.

An **OD system** takes a "Hot Water Priority" approach. If the GHP is not in a heating or cooling mode, or is turned off, the OD turns on the compressor and uses it only for hot water. If the GHP system is running and there is a demand for hot water, the OD diverts heat energy for hot water. Its fast response has virtually no impact on home heating. It is a primary domestic hot water source with its own storage tank and does not require electricity or gas auxiliary backup. The OD provides 100% of domestic hot water needs. In 2010 it added $2,250 to the system cost.

OD plus Desuperheater: And finally, HydroHeat offers a system that combines the desuperheater and the OD capability into a single heat pump, providing both Hot Water and Home Heating Priority simultaneously. ✺

Driveway Snow Melting Keeping a driveway free of snow may be practical in some commercial applications where there is a great amount of refrigeration equipment that continually produces waste heat. But for residential use, it may require a heat pump larger than that needed for an average home. For more information, see "An Informational Survival Kit" at *www.HeatSpring.com*

Chapter 11

Geothermal Plus Solar Energy: A Perfect Match

If there were ever a marriage made in heaven, it would be to pair a geothermal heat pump with a photovoltaic solar-electric installation. They both use nature as an energy source. The GHP may be free of fuel costs, but it does add electric costs when running. If you are connected to the electric grid, a solar photovoltaic (PV) system can reduce your electrical load because every watt-hour your solar system delivers is a watt-hour you do not have to buy from your electric utility company. And this is true whenever the sun is shining because sunlight is converted into free electrical energy.

The beauty of a PV installation is that when it is connected to the electrical grid, the grid is like an infinite, free storage battery. Solar power cannot only feed electrical energy into your home to reduce the amount needed from the electric company, but during those times when the solar system generates more power than you are consuming, it sells electric power back to the electric company. It does this by automatically reversing the electric meter, spinning it backwards as the excess power is diverted to the utility grid. If your utility offers net metering, your electric bill will be lowered and you will get a "credit" at the same rate you pay them.

Not all states have laws that permit net metering although many utilities provide it anyway. One alternative is to use a second meter to measure the power going to the utility, but then you usually get reimbursed at a much lower rate. The Database of State Incentives for Renewable Energy has a website *(www.dsireusa.org)* that lists a wealth of information, including which states mandate net-metering.

There are only three primary components that are wired together in a grid-tie system: an array of silicon photovoltaic solar panels mounted on an aluminum structure(s), an inverter to convert the direct current (DC) out of the array to alternating current (AC) that is usable in the home, and a meter supplied by the utility company. (To learn more about your grid-tie solar/wind options, read *Got Sun? Go Solar, 2nd Edition* by Rex Ewing and Doug Pratt.)

With this setup, if the grid has a power failure, you have a power failure. A propane-fired emergency power generator is one solution. Battery backup for selected circuits is another.

Roof-mounted solar arrays are ideal for homes with south facing roofs that do not have large trees that would cause shading.

A home on Long Island, New York, with a 2.7 kW array of grid-connected PV modules.

PHOTO: EVERGREEN SOLAR

PV panels can be mounted on flat or pitched roofs. Ground- or pole-mounted systems are also effective installation alternatives.

A typical small home may have a 450 square-foot (42 sq m) south-facing roof installation that can fit 30 PV panels that would,

Snow is easily removed from ground-mounted solar arrays, so you can keep charging with the sun. PHOTO: LAVONNE EWING

depending on climate and latitude, average 16 kilowatts-hours (kWh) per day of DC power. This would typically provide 6,000 kWh over a year; at $.15/kWh, that equals about $900 in savings annually. In my case, 6,000 kilowatt-hours would cover my geothermal power needs, but I would need a far bigger solar system to cover my total annual costs. Some analysts assume that electrical costs will increase about 5% each year so generating your own electricity will become more important in the future.

But then we must consider the total installation cost. A 5-kilowatt system as above might cost around $42,000 in 2010. In New York, a state rebate will pay $20,000 of that. Federal and state tax credits will total $7,000. Total costs drops to around $15,000. At $900 savings per year, that makes it a 16.7-year

A typical grid-tie installation with an inverter in the center (cover is off for wiring), a DC disconnect on the right, and an AC disconnect on the left. Far left is the existing house meter and main breaker panel, also with the cover removed. PHOTO: DOUG PRATT

payback. Another avenue gaining popularity: leasing a PV system to eliminate upfront costs. *FindSolar.com* lists installers in your area.

All of this tracks perfectly with a new trend (actually, an old concept with new clothing) where the emphasis is on local energy solutions. As reported in *Scientific American*, it makes more financial sense to generate new power closer to home. Wind turbines and rooftop solar panels could provide 81% of New York's power, according to the Institute for Local Self-Reliance *(www.ilsr.org)*.

Local energy is not a romantic notion. Plans for the world's largest wind farm in Texas had the plug pulled because the transmission lines were too expensive. Instead, a series of smaller installations closer to major cities is planned. More and more companies are installing "micropower" plants to power a building or campus. Germany stopped subsidizing the fossil-fuel industry and enacted policies to force the utilities to buy power from local people at a decent price. They now have 1.3 million photovoltaic installations, more than any country on earth. ✪

Geothermal for Off-Gridders?

Not everyone wants or can be connected to the grid. In those off-grid cases, they would need a few more components and a battery bank for storage and backup. You should be aware, however, that GHPs use more electricity than most off-grid solar/wind systems are designed to provide, particularly in winter when the sun's output is typically 40–50% less than in summer. If you are contemplating installing a GHP in an off-grid system, be sure to discuss it thoroughly with a seasoned solar installer before setting your plans in stone. You may just discover that the cost of solar panels and lead-acid batteries is just too high to be practical.

Ukiah, California

Doug Pratt's home in Ukiah, California, is a perfect example of marrying GHP technology with solar technology to obtain real synergy. His GHP provides four units of heat energy for every unit of electric energy to run it. "Almost like stealing," as Doug puts it. And his solar PV installation then pays for that one unit (and more) of electrical energy cost. His solar electric system delivers about 20 kWh per day when the sun is shining, and over time practically wipes out his total electric bill by running his meter backwards. He has an all-electric, passive solar home, which means that its energy-efficient design takes full advantage of the sun's energy, reducing the need for grid power. His annual electric bill: $110. I wish I could say that.

His active solar installation has 48 peel-and-stick solar panels located on the roof of his workshop; sealed batteries store the energy and two 3,600-watt inverters convert the DC solar power to usable AC power. Peel-and-stick modules are made of amorphous silicon that require about 50% more surface installation

Earth Smart Home

area to get the same amount of wattage as crystalline silicon, but can be built without glass covers for an unbreakable module.

His one-story home includes 1,450 square feet (42 sq m) with a single-zone, 2.4 ton GHP (WaterFurnace brand) connected to vertical closed loops. It has two 250-foot (76 m) deep boreholes to provide heat and cooling. The best thing about his setup? "I can make the fuel," he said. Well put, I say. His total GHP cost was $20,000 of which $12,000 was for the backyard hole drilling. That is $24 per foot, about standard.

Somehow, I cannot feel sorry for him when his biggest disappointment in the GHP is that it is so quiet that he can't tell if it is running without sticking his head in the garage. In the garage? I keep forgetting that California homes do not seem to have basements.

The garage roof holds the peel-&-stick PV panels; the inverters, charge controllers and batteries are installed inside. PHOTOS: DOUG PRATT

Chapter 12

What Will It Cost?

There's one other difference between a refrigerator and a GHP. You can get an instant price on a refrigerator, but coming up with a price for a GHP system requires an experienced and qualified contractor because there are so many variables.

First of all, the contractor will use a computer program to determine the proper size of the GHP unit—the refrigerant loop. That size is determined by the design heat loss for that home or building. The size of the GHP must be large enough to replace the maximum heat that is being lost by the home during the coldest weather. It is usually expressed in tons. A ton is equal to 12,000 Btu of heat energy per hour. A Btu is the quantity of heat needed to raise the temperature of one pound of water by one degree Fahrenheit. All total costs flow from that initial calculation of tons of heat energy needed for your home. Most modest-sized homes require from 3 to 5 tons.

Once the size (in tons) is determined, the ground loop can be designed, again using a computer program, to transfer the proper amount of heat energy from the ground to the refrigerant loop. The size and type of that ground loop heavily impacts the cost.

The final cost segment comes from choosing the method of getting the heat into the living areas of your home, such as forced air or hot water. But in order to get some general cost guidelines before actual quotes are requested, here are some suggestions.

Using an Online Calculator

You can get a rough idea now of what a closed-loop system would cost by checking a website cost calculator where you plug in details about your home. Try: *www.geosunnrg.com/geothermal-cost-estimator*. GEO SUN NRG provides professional ground-loop engineering services and also financial solutions to eliminate the upfront installation costs. In 2010 I used their calculator for my home as if I were a new customer with new construction using the closed-loop approach and got the following reply:

> **Installation Cost Estimate: 4-ton Closed-Loop System**
> *If you have an area of 150 feet by 40 feet, or 6,000 square feet, unencumbered by driveways, outbuildings, landscaping, septic systems, etc., you may be able to install a **horizontal system** for $21,000 to $27,000 ($5,250 to $6,750 per ton).*
>
> *If your property will not accommodate a horizontal system, you will require a **vertical system**. Installation cost range: $28,000 to $34,000 ($7,000 to $8,500 per ton).*

The actual costs for our open-loop system was $24,500 in 2007 and that included $5,500 for ductwork. That comes to about $6,100 per ton.

The EPA has stated that geothermal systems increase the value of a home $20 for every dollar saved in oil/gas heating costs per year.

Average Cost Breakdown by Segment

Another way to look at this is to consider segments:

Hardware for a 4-ton, 2-zone, 2,400 square-foot (223 sq m) home, including GHP hardware, transportation, desuperheater, thermostats, hot water heater: about $9,000–$13,000 (in 2010).

Labor: about $3,500–$5,000.

Heat/Cooling Delivery: Any system—oil, gas or GHP—can deliver heat into the home whether it is hot-water heat (baseboard radiators or in-floor radiant) or hot air. But ducting air is the only way to get both heating and cooling. Insulated ductwork: $3,000–$6,000.

Ground Source Options (listed by increasing costs):
- **Open-loop well-source piping** from well to house and to GHP and return (excludes well costs): $2,000–$3,000.
- **Closed-loop pond or lake source**: $2,000–$4,000 (distance to the water is the variable factor).
- **Closed-loop horizontal trenching** using "slinky" coil installation: $4,000–$8,000.
- **Closed-loop vertical installation** can cost from $15,000 to $25,000 for the drilling and piping. However, it requires far less real estate. Drilling runs about $13–$15 per foot plus inserting, connecting and sealing the pipe runs about $8–$10 per foot. Actual 2010 examples of two typical systems in New York State: Three 250-foot (76-m) holes cost about $16,000 while two 600-foot (183-m) wells cost about $25,000 (each averaging about $21 per foot).

For another perspective on costs, I have included data from a 2008 study by Kevin Rafferty, PE, distributed by the HeatSpring Learning Institute *(Figure 12-1)*. These are total installed costs for a 3-ton system including both inside costs (ductwork, heat pump, controls, etc.) and outside costs (ground loop, piping, excavation and/or drilling). The totals will add up to a high number, but this should be compared to the cost of a high-efficiency furnace or boiler plus a separate full-home AC system.

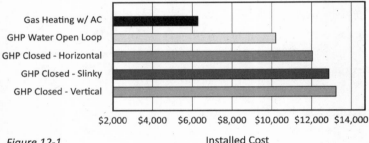

Figure 12-1

Source: "An Information Survival Kit", 2008, Kevin Rafferty, PE

Existing Ductwork?

If you add a shiny new GHP to an older home that already has ductwork installed, you could save a lot of money and work. Or will you? Older ductwork for forced air delivery systems was often designed to deliver air at high velocity. GHP metal ducts, on the other hand, are insulated and designed for a slower speed air delivery. This setup takes less electricity to run the blower, but it also means that larger ductwork is needed so that more air can be delivered. You should have your older ductwork inspected by your installer to determine if its size is usable and if it should be insulated.

DX Costing

Sub Terra Energy Systems of Newberg, Oregon, (*www.sub-terra.com*) offered in 2010 a fixed-price installation of a DX system. They only operate in Oregon and Washington, but the costs could be generally useful. However, these GHP systems are designed for forced-air delivery only and do not include the cost of the ductwork. If you already have ductwork installed, you'll get a good idea of the cost. If not, it could add another $3,000 to $6,000.

    3 Ton=$18,000 or $6,000/ton *(plus ductwork)*
    4 Ton=$20,500 or $5,125/ton *(plus ductwork)*
    5 Ton=$22,500 or $4,450/ton *(plus ductwork)*

In Summary

Providing a homeowner with an accurate cost for his/her home is beyond the scope of any book, but these cost figures will serve as helpful guidelines. Your actual installation time frame will be different from the examples shown; your home's design and construction will be unique as will the physical location and ground soil conditions; your local codes and labor costs will also vary. So when you receive those first bids from the contractors, the

---

With a conventional furnace or boiler, it is common practice to oversize the unit to compensate for its low efficiency and the home's poor insulation. **But with a GHP, it is a bad idea to either oversize or undersize the unit.** If too small, it will run continuously (wasting electricity) and never get the home warm enough. If too large, the cooling capacity would be larger than needed. This would cause frequent on/off cycling ("short-cycling") that causes excessive wear and reduces the ability to lower humidity. On the other hand, oversizing the ground loop is good.

odds are fairly high that the real numbers will give you "sticker shock." Do not despair. Consider the following facts:

In your new home, you are going to install and pay for some type of heating and cooling system anyway. The real question is, "What is the added cost of the GHP?" Then factor in the current US federal tax credit of 30% for very substantial savings, right from the start. And even if the added cost of a GHP is not completely offset with the tax credit, you will be ahead of the game because you will be saving thousands of dollars *every* year in zero fuel costs.

Keep reading—the next two chapters will explain the tax credit and payback issues in more detail. ✥

Novel Approach to Geothermal by Electric Cooperative

The Lake Region Electric Cooperative (LREC) with 25,000 members in Minnesota takes the word "Cooperative" seriously. They have initiated a unique business plan (called the EarthWISE Geothermal Program) that eliminates the initial costs of geothermal underground loop systems for homeowners who install geothermal heat pumps. For each home under this plan, LREC will custom design (using LoopLink software), install, and pay for the underground loop system. They would then charge a flat monthly fee (loop tariff) for a 35-year lease of the loop system. Alternatively, the homeowner can buy the loop at any time.

LREC does their own installation (up to the house), using horizontal directional drilling that requires less ground disruption.

By lowering the cost of installation, using local and federal incentives, and by reducing annual heating costs, a GHP system becomes very affordable. A solid proponent of geothermal ground source energy, LREC has also installed their own GHP system in their Operations Center.

Chapter 13

Taking Advantage of Rebates & Tax Credits

The time has never been better for homeowners to take advantage of green energy rebates and tax incentives. In addition to personal monetary benefits, these renewable energy subsidies support national objectives. The EPA has estimated that the United States alone could reduce its dependence on foreign oil by 21.5 million barrels of crude oil each year for *every* 1 million homes using geoexchange (as they call it) and reduce greenhouse gases by eliminating 5.8 million metric tons of CO_2 emissions annually.

USA

In April 2010, the EPA announced new, more rigorous guidelines for earning the Energy Star label *(www.energystar.gov)*. This includes higher efficiency standards for geothermal heat pumps and included home-based solar and wind energy installations. As a part of this, the EPA also issued tax credit guidelines to help the homeowner. In order to obtain tax credits for geothermal installations, the homeowner must prove that these standards have been met and installed. See the US Energy Efficiency Tax Credit Summary on the next page.

A listing of state subsidies and rebates can be found on Geothermal Heat Pump Consortium's website: *www.geoexchange.org*. New York provided a $500 cash rebate in 2007. Compare this to a photovoltaic solar installation for which New York provides up to a $20,000 payment. Another excellent website is the Database of Incentives for Renewables and Efficiency: *www.dsireusa.org*.

Canada

Canada's EcoEnergy Retrofit program, started in 2007, provided grants to eligible homeowners who improve the energy efficiency of their homes, including GHP installations. The homeowner had to arrange for a Natural Resources certified advisor to conduct an onsite assessment of the home's energy use. However, in April 2010 the Canadian government abruptly ended this program and was not accepting new applications. This sudden cost-cutting measure resulted in a large drop in 2010 GHP Canadian sales, and even US sales to Canadians.

There is hope that the public will recognize provincial subsidies where they are offered and that the market will come back. Only some of the provinces offer help in the form of rebates and loans. In Prince Edward Island, a retroactive tax exemption is provided. To learn more, go to: *http://oee.nrcan.gc.ca* and click on Grants and Incentives to keep up to date on federal incentives. For the latest information on provincial subsidies, go to *www.nextenergy.ca*. ✺

If just one in 12 California homes installed a geothermal system, the energy saved would equal the output of nine new power plants.

US Energy Efficiency Tax Credit Summary

The US federal tax credit of 30% with no cap is a very powerful incentive to invest in renewable energy. Keep in mind that it will be applied to your federal tax return the following year. So if you owe $12,000 in taxes but claim an $8,000 energy tax credit, your tax liability will be only $4,000.

In order to obtain tax credits, the homeowner must do a small bit of work to prove that a qualified installation has been completed. Here is how:

1) Save your bid, bills and receipts.
2) If applicable, obtain a manufacturer's Certification Statement for each product (it should include the efficiency rating of your specific model number). This can usually be done online, or it may be provided with your manual.
3) File IRS tax Form 5695 with your annual tax return.

▶ **Tax Credit: 30% of total cost**
▶ **Expires: December 31, 2016**

The following details apply to systems installed after 12/31/08.

The best place to find more information: ***www.dsireusa.org***

Geothermal Heat Pump Efficiency Requirements
See Chapter 16 for definitions of EER and COP
▶ Closed-Loop Efficiencies: EER ≥ 14.1 and COP ≥ 3.3
▶ Open-Loop Efficiencies: EER ≥ 16.2 and COP ≥ 3.6
▶ Direct Expansion (DX) Efficiencies: EER ≥ 15 and COP ≥ 3.5
▶ Federal tax credit includes installation costs for existing homes and new construction. The home served by the system does *not* have to be the taxpayer's principal residence.

continued

Small Wind Turbines
- Federal tax credit includes installation costs for existing homes and new construction. The home served by the system does *not* have to be the taxpayer's principal residence.

Solar Electricity
- Federal tax credit includes installation costs for existing homes and new construction. The home served by the system does *not* have to be the taxpayer's principal residence.
- Photovoltaic systems must provide electricity for the residence and meet applicable fire and electrical codes.

Solar Water Heating
- Equipment must be certified for performance by the Solar Rating Certification Corporation (SRCC) or a comparable entity endorsed by the government of the state in which the property is installed.
- The home served by the system does *not* have to be the taxpayer's principal residence. At least half the energy used to heat the dwelling's water must be from solar.
- Does not apply to swimming pool or hot tub heating.

Established by the federal Energy Policy Act of 2005, the federal tax credit for residential energy property initially applied to solar-electric systems, solar water heating systems and fuel cells. The Energy Improvement and Extension Act of 2008 (H.R. 1424) extended the tax credit to small wind-energy systems and geothermal heat pumps, effective January 1, 2008. Other key revisions included an eight-year extension of the credit to December 31, 2016, the ability to take the credit against the alternative minimum tax, and the removal of the $2,000 credit limit for solar-electric systems beginning in 2009. The credit was further enhanced in February 2009 by The American Recovery and Reinvestment Act of 2009 (H.R. 1: Div. B, Sec. 1122, p. 46), which removed the maximum credit amount for all eligible technologies (except fuel cells) placed in service after 2008.

Chapter 14

Payback: When Do You Get Your Money Back?

This may be the most important chapter in this book. Why? Because the decision to install a GHP in your home, to expend or borrow scarce funds, will largely depend on whether or not you believe you will get a full return on your investment.

And so I thought about this very carefully because I recognize my pro-geothermal bias. After all, my own installation in my new home had an immediate positive payback according to my calculation. But that may not be everyone's calculation.

First, I need to define what we are talking about. ROI (Return on Investment) is a banker's term that can dull senses. I like "payback" better. Payback is a term of time (years usually) when you finally get your money back. From that time on, you are ahead every year. You are spending money now to save money later.

The usual calculation for payback compares the *added* costs of a GHP over and above the cost for an oil, gas or electric heating/cooling system, also taking into account the annual savings that comes from eliminating the cost of fuel needed to run a furnace and an air conditioner.

To my simple mind, payback is very clear. You are going to

pay for and install *some* type of heating system. If a GHP costs more, but offers greater savings over time, you just need to know when you'll break even for the extra cost.

As I've said before, GHP manufacturers have difficulty in providing a payback for a specific home beforehand because of all the variables—GHP cost, oil consumption at your home, prevailing fuel costs, ductwork costs, etc. Many provide a general statement: "We often see a three- to five-year payback of the additional costs." Some companies provide a handy savings calculator where you can plug in your data and get a savings calculation. Northern Heat Pump in Canada actually has an ROI calculator. Try one or two of these calculators.

That may be enough for many, but you cannot see *how* they compute results or what factors they include. There are several ways to calculate your return on investment, and we'll take a look at a few models. Pick the one that fits best; then plug in your own numbers.

Model 1—Payback on Added Costs, No Subsidy

You spend $28,000 for a GHP system; $8,000 more than the $20,000 cost of a natural gas boiler and full AC. You will save $2,500 every year by not buying gas, but it takes $200 of electricity per year to run the GHP. So what is the payback?   3.5 years.

   $2,500 - $200= $2,300 savings per year
   $8,000 ÷ $2,300= 3.5 year payback

Model 2—Payback on Added Costs, With Subsidy

The above model ignores any subsidy or tax credit that may be available. I will use the current US 30% federal tax credit, but in Canada you could insert your provincial rebate or credit. As you will see, adding the 30% tax credit makes a big difference, as it should in order to accomplish its purpose of encouraging the use of renewable energy.

> *$28,000 GHP cost x .30 = $8,400 tax credit*
> *$28,000 – $8,400 = $19,600 actual cost*

The payback? Zero years! When compared to a $20,000 natural gas system, the payback is immediate. Think about it—the GHP costs no more than the natural gas system.

In five years you will have saved another $11,500 in fuel savings and you did not even have to install a chimney! And we are ignoring the fact that any furnace/burner also requires a bit of electrical power as well as annual maintenance costs; they also have shorter life spans.

Model 3—Payback with Full Financing and Subsidy

Take the same example as above but now also consider the cost of financing the added GHP costs. A tax credit would not be available until after April 15 of the following year and you have planned all along to include the total heating system cost within your construction mortgage and use the tax savings the following year for a photovoltaic solar panel installation. In effect, one tax credit finances the installation of another renewable energy source. Smart move!

In folding the additional $8,000 into your home mortgage, your mortgage payments increase slightly to cover the added principal and interest. Assume that the additional charges add up to $500 per year. So here's how it breaks down:

> $2,500 annual fuel savings
> − $200 added electricity
> − $500 extra mortgage costs
> = $1,800 savings per year

The GHP incurs no additional cost because your total net savings starting that very first year is $1,800.

And your payback now? Zero years! You have immediate payback even without applying the subsidy. And in ten years you will have saved $18,000.

Model 4—Payback on Total GHP Costs

Assume your total GHP system (including installation) is $30,000 and your net fuel savings per year are $1,900. Your federal tax credit will be $9,000 (30% of $30,000). A standard oil furnace with full AC would cost $22,000. What is the payback on the entire investment, not just the added cost? 11 years.

> $30,000 − $9,000= $21,000 system cost
> $21,000 ÷ $1,900= 11-year payback

This is the most conservative approach imaginable. It means that the savings must cover the entire GHP cost! There are some who think that this is the correct approach. But they don't seem to realize that if you applied this payback concept to the alternate $22,000 system, the payback would be an infinite number of years—there are no savings.

> $22,000 ÷ zero =∝

Furthermore, the net cost of a GHP was actually less than the oil burner because of the subsidy. Thinking long term, in

20 years the savings will have paid for the cost of the GHP plus $17,000 more.

$1,900 x 20 = $38,000
$38,000 − $21,000 = $17,000

Model 5—Payback on Existing Home Renovation

You are replacing your home's obsolescent oil burner with a GHP. You need both air delivery (for air conditioning) and hot water delivery for your existing radiant floor heat. You paid $4,000 the previous year for heating fuel. A new high-performance boiler and separate AC would cost $16,000 if installed. Final GHP costs are $32,000 because of the vertical boreholes for a closed-loop system. You receive a tax credit the following April of $9,600. The cost to operate the GHP is $300 per year. What is your payback? 1.7 years.

$32,000 − $9,600 tax credit = $22,400 GHP cost
$22,400 − $16,000 = $6,400 added cost of GHP
$6,400 ÷ $3,700 yearly operating costs savings = 1.7 years

In 20 years you will have saved over $74,000 if oil prices do not go up and that is highly unlikely. The rate of return on investing $6,400 with an annual return of $3,700 for 20 years is 58%!

If you think long term, you will see that a GHP will save you a lot of money. The added cost will be more than repaid and there is no pollution. Is there any better way to invest money with such a high return coupled with such a low risk? Geothermal heat pumps are the answer. ◉

Geothermal Savings Calculators

How much money can you save by installing a geothermal heat pump? Many GHP manufacturers provide savings calculators on their websites. Just answer a few questions about your home, your existing heating/cooling system and hot water needs, such as:

- Your nearest city/state
- New or existing home
- Square footage to be heated/cooled
- Insulation/air leakage: poor, average, good or excellent
- HVAC equipment age: 0–3 years, 3–15 years, 15+ years
- HVAC equipment efficiency: standard or high efficiency
- Heating type: forced air, radiant floor
- Home heating fuel: natural gas, electric heat pump, electric resistance, propane, fuel oil
- Water-heating fuel: natural gas, electric, propane, fuel oil
- Energy efficiency improvements, such as insulation upgrades, Energy Star appliances, compact fluorescent lighting
- Ground area available for GHP loop: horizontal, vertical

Feedback includes energy costs savings by switching to geothermal, where your energy dollars are spent (for heating, cooling, hot water, appliances, lighting), your carbon footprint and more.

Also provided by some websites: Design Data for these calculations (outdoor design temp, heating degree days, earth temp, current building load, hot water usage) and Utility Rates for summer and winter.

Some of the GHP manufacturers with online calculators:
- Climate Master *www.climatemaster.com*
- FHP Manufacturing *www.fhp-mfg.com*
- GeoComfort *www.geocomfort.com*
- Hydron Module *www.hydronmodule.com*
- Northern Heat Pump *www.northernheatpump.com*
- TETCO *www.tetco-geo.com*
- WaterFurnace *www.waterfurnace.com*

Chapter 15

Finding a Contractor

A heat pump installation requires a higher level of knowledge and experience than an oil or gas burner installation. The GHP is simple in concept but complex in practice. This is not a do-it-yourself project. It has feedback loops, safety controls, fault sensors, shutoff circuits, diagnostic LEDs, and a large microprocessor control board. (Amber LED S-12 is blinking—what does that mean?) You do not plug a GHP into a wall socket. The contractor must consider the entire system, from the heat source to the final comfort of the occupants. So make certain that your contractor is qualified and experienced. GHP contractors will also install heating, ventilation and air conditioning (HVAC), but not all HVAC contractors are experienced with GHP systems; they are fewer in number and they are getting busier.

Your goal is to find an experienced, qualified local geothermal contractor and a high quality geothermal manufacturing company with the latest technology. There are two ways to do this. You can search locally for a contractor/installer and use his judgment and experience in finding a manufacturer, or you can select a manufacturer and use their website to find a qualified, accredited contractor.

Searching for a GHP Manufacturer

If you are searching for a manufacturer, review the appendix list of US and Canadian manufacturers and then make sure they:

- **Have qualified the hardware** to US EPA (Environmental Protection Agency) Energy Star standards, or ISO 13256 (International Organization for Standards) certification, or Canadian CSA C13256 (Canadian Standards Association) standards. See *Chapter 18 – The Importance of GHP Performance Standards* for more information on standards.
- **Provide the latest technology** in high-efficiency compressors (two-stage, digital scroll compressors, etc.).
- **Offer a solid, comprehensive warranty.** The appendix lists residential GHP manufacturers and their warranties. It shows wide variations in warranties, from one year to 20 years on different parts. Warranties make us feel good, and if the system fails (it will always happen in a severe snowstorm), they will become very important. But this should not be the sole criterion used. For one thing, if the system survives the first few years, it will probably keep working for 20–25 years. Service and GHP efficiency are also very important. I would be concerned, however, if no warranty is listed. For a baseline warranty, you could use the former EPA Energy Star requirement for their Energy Star Partners

---

By using underground thermal sensors, Sub Terra *(www.sub-terra.com)* has determined that most of the heat transfer takes place within a 2-foot radius around each DX tube. This ***area of influence*** collectively becomes a reservoir of heat capacity. For their DX installations, they keep a minimum of 4 feet (1.2 m) between tubes to avoid robbing heat from one another and to prevent the ground from freezing.

warranty: two years on parts and labor; five years parts and labor on the refrigerant loop (including the compressor). Some companies resist this, especially the labor.
- **Have good efficiency ratings.** COP ratings (Coefficient of Performance) for heating should be:
    Closed-loop with Hot Air ≥ 3.5
    Closed-loop with Hot Water ≥ 3.0
    Open Loop with Hot Air ≥ 3.8
    Open Loop with Hot Water ≥ 3.4
    DX ≥ 3.5
- **List the names of qualified contractors in your area.**

Searching Locally for a Contractor/Installer

If searching online or by telephone directory for a local contractor/installer, consider that the International Ground Source Heat Pump Association (IGSHPA) has a list of accredited contractors, listed by state or Canadian province: *www.igshpa.okstate.edu*. When you find a contractor, inquire about the following:
- Have they installed a GHP in their own home?
- Are they an accredited installer/contractor?
- How many GHP systems have they installed, both open loop and closed loop?
- Do they use load-calculation software to determine the correct size of unit for your home and the loop field needed? *(This calculation is very important—insist on it. The time has passed when back of the envelope estimates are good enough.)*
- Do they have trenching equipment or do they work with a capable trenching or well-drilling subcontractor with geothermal drilling experience?
- Will they provide a written statement of cost, delivery and warranty?

- What warranty do they offer? Even more important—what warranty does the contractor offer on his installation? *Get this in writing. Save all this for tax purposes.*
- Which manufacturer's equipment are they using? How long have they used this equipment? What does he/she like about that manufacturer?

A three-day accreditation course is a necessary but insufficient measure of installation expertise. Trainer Michael Hunt has found that he has to work with 90% of the class participants for a year or two after the class to get them on the right path. There is no substitute for experience and mentoring. A contractor in your area is going to be better at sizing and designing the complete system than one many miles away without geological knowledge of the area. If your installation requires drilling boreholes, the contractor should either have his own drilling rig or subcontract for it.

Other Questions to Ask

You can also ask questions about the more efficient DX systems vs. standard approaches (keeping in mind that DX installers are not as common as those experienced in standard GHP systems). Ask about desuperheaters and about tankless, "on demand" water heaters. What about slinky loops? Ask about the payback, efficiency, operating costs and system life span.

Get references. Call them. This cannot be understated.

Your contractor should run software programs to determine the size of the unit needed and then the amount of ground piping to obtain the ground energy needed. This involves the house size, tightness, insulation, windows and number of zones to calculate the design heat loss. The analysis should also include the degree-days, the winter and summer inside design (preferred) temperatures, the winter and summer outside average temperatures, geology of the area, type of ground, size of land area and even the elevation. What works in your backyard will probably be different in Winnipeg or spots of Wyoming.

In my case, it worked out to about one ton of capacity for each 600 square feet (56 sq m) of house. Our 2,400 square-foot home (223 sq m) required a 4-ton unit and that has proven perfectly adequate for our land and weather. My elevation is only 249 feet (76 m) above sea level, although I can see the 4,000-foot Catskill Mountains across the Hudson River. Move my home anywhere else and the unit size may be higher or lower.

Some contractors discourage the use of open-loop systems because the water is less controllable and local codes must be followed. Listen to them, but consider the lower installation cost benefits. Other contractors will look for open-loop possibilities as a first choice, since that is the most efficient and the least costly to install. They will also determine if there is real estate available for a horizontal ground loop, and finally a vertical hole. It is very important to not oversize the unit by installing one too large for the load in your home. If the GHP ton-rating is higher than needed, it could start a "short cycle" where frequent on/off cycling of the cooling system would occur as it reaches the thermostat's setting too quickly. This can shorten the life of the unit. Oversizing the ground loop, on the other hand, can be an advantage especially if there are plans to increase the house size in the future.

Shallow ground temperatures are relatively constant on this continent, therefore geothermal heat pumps can be used effectively almost anywhere. But determining the best type of ground loop suitable for the composition and properties of your soil and rock require experience. Soil with good heat transfer properties (i.e. soil with moisture) requires less piping to gather the heat energy needed. If extensive rock ledge is present, vertical ground loops may be required. Ground or surface water availability may be suitable for an open- or a closed-loop system. You need a contractor familiar with the local topography.

Geothermal systems can be installed in new or retrofit applications. It could be less costly in a retrofit because you may have the hot air ductwork or hot water delivery system already installed. But if you are building a new home, you are an especially good candidate. You are already making a long-term investment with the mortgage so the initial cost of the GHP can be tied into your monthly payment. Your savings in fuel costs will generally more than cover the additional amount added to your monthly mortgage payment. So you are creating a positive cash flow or payback—right away! ✺

Specialized high-speed, lightweight drilling rigs are often used to punch the vertical bore holes. Here you see a loop pipe being inserted into a bore hole. PHOTOS: DOUG PRATT

Determining the Size of Your GHP System

To determine the size of a geothermal heat pump system (in tons) as well as the size of the loop field, the heating/cooling loads (heat loss/heat gain) of that specific residence must be calculated. To do this, a great deal of information is needed about the home and its location. The homeowner can assist by having some of this data available ahead of time. Examples of the type of data needed for plugging into the software:

- Location, elevation, latitude
- Average wind speeds
- Indoor temperature settings for owner's comfort
- Local outdoor temperature/humidity ranges
- Which side of house faces north
- Window/door placement, material, low-e value, double- or triple-glased, and size
- New construction vs retrofit
- House plans for size, zones, stories, insulation, tightness, ceiling heights, etc.
- Basement—finished or not; heated and/or cooled
- Garage—heated?
- Number of fireplaces
- Heat distribution system (forced air, radiant heat)
- Soil type and moisture content (for horizontal)
- Water-well log data, if nearby (for vertical bore site)
- Site plan: gardens, walls, poles, outbuildings, landscaping, available ground area, etc.

"Installation of the ground field is the most critical aspect of a GHP system, whether it is DX or water based." Curt Jungwirth, 35-year GHP expert

Town and Country, Missouri

Earth Smart Home

This large, 10,000-square foot (929 sq m) new residence in the city of Town and Country, Missouri, near St. Louis, is unusual because in addition to providing geothermal home heating and cooling, the contractor has carefully engineered a separate geothermal domestic water system that provides unlimited hot water for household use as well as the swimming pool.

Hoffman Brothers – St. Louis Geothermal Heating and Cooling Contractors, who designed and installed this system, are not just HVAC installers; they have experienced mechanical engineers on staff to create customized systems. For this home, they installed GeoComfort heat pumps totaling 21 tons of capacity for the home heating and cooling plus a separate 6-ton unit for water heating. The ground source for all of these units is 21 vertical borings, each 150 feet (46 m) deep, using thermal grout in a closed-loop design.

The separate water-to-water heat pump provides hot water to two 80-gallon (303-liter) storage tanks plus a circulating line to keep the lines hot at all times. This eliminates any delay in delivering hot water for showers, etc. In addition, this unit heats the pool at all times. Beyond this, there is a backup on-demand (tankless) water heater so that all hot water needs can be met if needed at the same time.

The total cost was $220,000 in 2009. A $66,000 tax credit gave them a net cost of $154,000. Energy savings are estimated to be $10,000 per year. When considering a standard system would've cost $92,000, they have a payback of only 6.2 years.

Part II

Unraveling the Science and Technology of Geothermal Heat Pumps

Chapter 16:

The Inside Story of Geothermal Super Efficiencies

Thunder and lightning are two spectacular and observable phenomena caused by of the movement of heat energy, increases/decreases in pressure and temperature, and the transformation of one form of energy to other forms.

In an analogous manner, the ability of a heat pump to create temperatures high enough to heat your home and efficiencies high enough to save money are measurable results of the manifestations of the same laws of thermodynamics as those governing planetary weather effects. But in a GHP, it all happens in a closed system which is hidden from view.

The purpose of Part II is to unravel this geothermal technology by providing a peek inside a heat pump—to see what is going on and why it really works. Opening that door should lead to some interesting insights.

In giving local talks, and even in after-dinner conversations about heat pumps, two questions keep popping up:
1) How can a GHP get efficiencies over 100%? Is that not getting something for nothing? This chapter will address that question.

2) If you put your hand in 50°F (10°C) water, it feels really cold. How can anything that cold provide so much heat for your home? Chapter 17 will respond to that question.

First, to review: heat is a form of energy. Energy is defined as the ability to do work and comes in many forms: electrical, mechanical, heat, nuclear, light, etc. Measures of heat energy include the calorie, joule and Btu (British thermal unit). Measures of electrical energy include the kilowatt-hour (kWh).

The Laws of Thermodynamics, in particular the First Law, tell us that energy cannot be created, destroyed, used up, dissipated or made to disappear. However, it can and will be changed into different forms and states. Nature does this automatically; we can show it mathematically. For example, to convert kWh into Btu, we merely multiply kWh by 3,412. There are 3,412 Btus of energy in one kilowatt-hour (kWh).

First Law of Thermodynamics

The First Law of Thermodyamics is an expression of the Law of Conservation of Energy, an empirical law of physics. It states that the total amount of energy will remain constant over time. As a consequence, energy cannot be created or destroyed, only transferred into different forms. In some cases, however, it can be stored. A pertinent example of energy being transferred is friction. Electrical energy is converted into mechanical energy to drive a motor. That motor has moving parts that rub, causing friction. Friction has then converted the mechanical energy into heat energy. Another consequence of this law is that the output of a thermal system cannot exceed the input. That means that the thermal efficiency of a device must never be greater than 100%. So how can a GHP obtain 400% efficiency?

Heat is different from temperature, but the two are closely linked; two sides of the same coin. Heat is a measure of how heat energy is transferred from one body to another. Temperature is the measure of the average kinetic energy in a body.

As energy changes state and is transferred into a body, the more rapidly the atoms move and the greater the temperature. In a closed-refrigerant system, such as within a GHP, temperature and pressure are directly linked. If you increase the pressure, the temperature will increase proportionately. This mandatory balance is embodied in the Ideal Gas Equation:

$$\frac{Pressure}{Temperature} = Constant$$

All of these fundamentals are at work inside the GHP.

Geothermal Super Efficiency

If you think that the efficiency of a heating/cooling system is an abstract concept that does not impact your pocketbook—think again. The higher the efficiency, the lower your operating costs. This factor gives a geothermal heat pump its high efficiency and the lowest cost per Btu of any other heating source. The oil guzzlers in our old schoolhouse could achieve a maximum of 86% efficiency by burning (changing the state of fuel oil into heat) and this was only after a fairly expensive annual tune-up. Where did that 14% go? Most of it went up the chimney as wasted heat energy and pollutants.

The efficiency of any machine is measured by the degree to which friction and other factors reduce the actual work output. (Actual work output divided by the amount of energy input equals the efficiency of the equipment.) When changing the state of a material by combustion, the output cannot equal or exceed the input or we would have perpetual motion. So the value will never reach 100%.

$$\text{Efficiency} = \frac{\text{OUTPUT}}{\text{INPUT}} < 100\%$$

It has no dimensions, just a percent, but because the input has a financial cost, it can also be expressed more realistically as:

$$\text{Efficiency} = \frac{\text{Energy You Actually Get}}{\text{Energy You Pay For}}$$

But in an apparent disagreement, the US Department of Energy has stated that ground-based geothermal systems will achieve *400 to 600% efficiency*. An efficiency of 500% means than the energy *out* is 5 times larger than the energy *in*. How is that possible? We are now back to the original question: How can a GHP obtain efficiencies of over 100%?

This really amounts to a difference in definition, since the word "efficiency," in the strict sense, is used differently for a heat pump. Although it is still output divided by input, heat pumps (in their heating mode) are measured by a Coefficient of Performance (COP). Since these devices are moving heat and not burning an energy source to get heat, the amount of heat they *move* can be greater than the actual input energy (electricity). Therefore, heat pumps are actually a far more efficient way of heating than simply converting fuel into heat as in a furnace. Figure 16-1 provides a graphic look at how super efficiency is obtained in a GHP.

That figure now provides the first look inside a heat pump.

It shows that energy harvested from the ground is "free" so it is only considered on the Output side and not in the Input side of the equation. The important point is the Output is the sum of the Input electrical energy *plus* the free heat energy from the ground.

Output = Input Electrical Energy + Ground Heat Energy

In addition, as a part of the COP definition, the Input electrical energy in kilowatt-hours (kWh) is converted to Btu (1 kWh = 3,412 Btu). Now, all parts are in the same units.

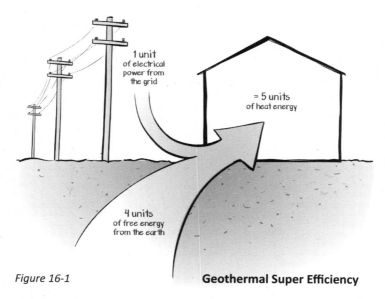

Figure 16-1   **Geothermal Super Efficiency**

Heating Efficiency = Coefficient of Performance (COP)

$$COP = \frac{OUTPUT \ (1 \ Input \ Unit + 4 \ Free \ Heat \ Energy \ Units)}{INPUT \ (1 \ Input \ Unit)} = 500\%$$

Reference: *Geothermal Heat Pumps: Guide for Planning and Installing* by Karl Oschner; Earthscan Publishing

Our graphic now illustrates a simple equation in a GHP heating mode:

$$COP = \frac{OUTPUT\ (Btu\ of\ Electric\ Energy + Btu\ of\ Ground\ Energy)}{INPUT\ (Btu\ of\ Electric\ Energy)}$$

The numerator (Output) must therefore always be greater than the denominator (Input) and in practice runs from about 3.5 (350%) to about 5 (500%), or even higher.

So, we have transformed the graphic into a simple COP equation, which in turn illustrates what is actually happening inside the GHP refrigerant loop. Now the mechanism inside the GHP that accomplishes this can be described.

Inside the Refrigerant Loop

Ground heat energy is introduced into the ground loop's refrigerant, which increases the temperature of that refrigerant and turns it into a gas. Then, using electricity, a compressor increases the pressure and that raises the temperature even more. Electrical energy has been transformed into mechanical energy.

Finally—and this is the critical part—the mechanical energy

What is a Refrigerant?

Refrigerant fluids are used in air conditioners and heat pumps to absorb heat energy at low temperatures and reject heat energy at higher temperatures. The HVAC industry has given trade names to refrigerants (which must be EPA certified) in order to identify different chemical makeups. Since R22 (Freon) has been linked to ozone depletion, it is now banned from manufacture. R410A (Puron) has become the standard replacement in this industry.

is then transformed into heat energy. Kilowatt-hours are being converted to Btu, just as in the equation, and the Input Btus are being added to the Ground Btus. So there are two sources of heat energy: a minor one from the transformed input energy and a substantial one from the ground loop with its pressure-enhanced temperature increase.

As far as cooling is concerned, the same process takes place except that instead of ground-heat energy plus converted-electrical energy being transferred *into the home*, it is house-heat energy plus converted-electrical energy being transferred back *into the ground*.

In the end, in accordance with the First Law of Thermodynamics as shown earlier, electrical energy has been transformed into heat energy and added to the ground-source heat energy to provide super efficiencies of well over 100%. You are getting something for nothing!

GHP Measurements of Efficiency

International, US and Canadian minimum efficiency standards have been established for heating in terms of the COP (Coefficient of Performance). In addition, standards are set for cooling, but defined slightly differently as an EER (Energy Efficiency Ratio). The higher the COP or EER, the more efficient the system is. Both ratings measure efficiency using Output divided by the Input, but in slightly different ways.

Heating: COP (Coefficient of Performance) is the total Output heating capacity in Btu divided by the electrical power Input, also converted to Btu. Numbers range generally from 3 to 5, which translate to 300% and 500%.

Cooling: EER (Energy Efficiency Ratio) is the total Output cooling capacity (in Btu/hour) divided by the power Input (in watts). Numbers can range from 9 to 31.

Homeowners must usually prove that they have met those efficiency standards in order to qualify for tax credits or rebates. They can obtain certification from the manufacturer's website or from their installer. My contractor left a copy with the other documentation that he provided.

GHPs are also rated by the ton. Most residential systems range from 2 to 5 tons. Mine is a 4-ton unit. A ton is a measure of the GHP's capacity to cool; it is the energy required to melt one ton of ice in one hour, which is 12,000 Btu per hour. (A Btu is defined as the amount of heat necessary to raise the temperature of one pound of water one degree Fahrenheit.) My system can move 48,000 Btu of heat energy per hour (at a specific earth-loop temperature) into my home. As the earth loop temperature changes, so does the output. That is why it is so critical to have a proper earth-loop design to achieve maximum efficiencies. For example: the cooler the earth loop, the higher the output and efficiency when in the cooling mode.

Figure 16-2 shows the Energy Star current minimum heat pump efficiency standards compared to the current best available and to my open-loop system. Note that these minimums will be

Comparison of Energy Star Standards vs. Real Systems						
System Type	Energy Star Minimum		Best Available		Author's GHP	
	EER	COP	EER	COP	EER	COP
Closed Loop	≥ 14.1	≥ 3.3	25.8	4.9	--	--
Open Loop*	≥ 16.2	≥ 3.6	31.1	5.5	19.0	4.1
*Water-to-Air System						

Figure 16-2

increased in 2013. Government standards and tough competition are encouraging COP and EER numbers that exceed the minimums and so we'll see actual operating efficiency increase over time until a peak is reached. High efficiency can be a criterion for product selection.

For additional details on Energy Star standards and efficiencies, see chapter 18. ✪

Eliminating COP/EER Confusion (Maybe)

Why are there two ways of measuring GHP efficiency in the US? In Europe, the term COP is used for both heating and cooling.

The COP (Coefficient of Performance) measures efficiency in the heating mode. The EER (Energy Efficiency Ratio) measures efficiency when operating in a cooling mode. What is really confusing is that the COP and the EER use different energy units for computing the efficiency numbers. The COP uses Btu as the measurement unit of energy in both the numerator (output) and the denominator (input). That is really handy because a result of 3.3, for example, means that the efficiency is 330%.

The EER, however, uses Btu/hour in the numerator (output) and watts in the denominator (input). The result is a much larger efficiency number. To put it in COP terms, we must divide the EER number by 3.412. That converts watts to Btu/hour. Instead of an EER of 13, you get 3.8 or 380% efficiency.

Why the difference? Possibly because EER numbers were designed after the SEER (Seasonal Energy Efficiency Rating) used for AC systems. Because GHPs *move* energy and because they are used year-round, they call it EER. In any case, when comparing companies, the higher the COP and EER, the better.

Mechanical Advantage Analogy: The Lever

From an energy standpoint, a GHP works in a similar fashion to ancient mechanical devices such as the lever, the inclined plane, the pulley, or the wedge. These simple devices have a mechanical advantage in the same way a GHP has an energy advantage.

A lever is one of those machines with an inherent mechanical advantage that allows a person to move a much heavier object than they would normally be able to. Archimedes stated the lever's mathematical principle in the 3rd century BC. "Give me the place to stand and I shall move the earth." In Egypt, the lever was used to lift and move 100-ton obelisks.

Figure 16-3

The ratio of the energy output of a mechanical system divided by the energy input is the measure of mechanical performance. Sound familiar? If the length of the lever to the right of the fulcrum is 3 feet (0.9 m) and to the left is 1 foot (0.3 m), then it has a mechanical advantage of 3.0. We can lift (output) 3 times (300%) as much as the input weight.

That is just what a GHP does!

Robert C. Webber, an American inventor, is credited with building the first heat pump during the late 1940s.

Chapter 17

A Look Inside: How a GHP Delivers Heat Into Your Home

In this chapter I will concentrate on why a heat pump is able to produce sufficiently high temperatures to heat and cool your home without the burning of fossil fuels and to answer the question: *"How does a GHP increase the temperature from 50°F (10°C) in the ground to 165°F (74°C) at the final heat exchanger?"*

The answer involves the transfer of heat energy and the conversion of one form of energy into another inside the GHP system while relying on the physical laws of nature.

I know from my own experience that a GHP is clearly able to heat my home very comfortably irrespective of the outside temperature. To do this in a water-to-air system, 165°F (74°C) is usually required at the final heat exchanger so that cool air can be heated, travel through the ductwork and enter rooms at about 100–105°F (38–41°C). In a water-to-water system, only 120°F (49°C) is needed to deliver radiant or baseboard heat.

I use 50°F (10°C) as an average ground temperature only as an example. Ground temperatures vary greatly depending on geography, climate conditions, and many other factors that impact

GHP design and installation. In this section any ground temperature could be used to show the basic theory of why things work.

There are two basic ways to increase the temperature of a substance in a closed system. First, when heat energy is introduced into a substance, its temperature will rise. If more than one source of heat energy is introduced to the same substance at the same time, they will each raise the temperature even further.

The second method is to increase the pressure. Liquids cannot be pressurized, but gases can be. Refrigerants are used because they can easily change state from a liquid to a gas and back to a liquid. Increasing the pressure will increase the temperature of a gaseous substance proportionately.

When we pry open the door to look inside a GHP, we'll see that both of these methods are at work. It all happens in a number of incremental steps to achieve that working temperature. I will outline those steps below and then apply the science behind each.

1) Earth Heat Energy

The first transfer of energy takes place when the sun's energy is absorbed by the earth, raising its temperature and keeping

---

The Third Law of Thermodynamics

The third law states that if all of the heat energy (thermal motion of molecules) were to be removed from a body, a state called absolute zero would occur. Absolute zero = 0° Kelvin or -469°F (-278°C). This is important because it demonstrates that even if the ground temperature is only about 50°F (10°C), it is over 500°F (260°C) above absolute zero and clearly contains a large amount of heat energy that is constantly being renewed.

it at a relatively constant level. Any substance at 50°F (10°C), even though it feels cold to your hand, actually contains a great deal of heat energy that can be transferred.

The field of thermodynamics studies the behavior of energy flow and as a result, physical laws have been established. The laws of thermodynamics describe some of the fundamental truths observed in our universe. The Third Law of Thermodynamics helps us to understand that any body even at 50°F contains heat energy. That is because it is over 500 degrees (F) above absolute zero, a theoretical condition in which no heat energy exists in a body. In other words, there is more than enough heat energy in the earth that can be continually captured and directed into the home.

2) Ease of Transfer

The next transfer of heat energy is from the earth to the ground loop of a geothermal system, raising the loop's water temperature. Then, in the next increment, that heat energy is transferred to the refrigerant loop, thus increasing its temperature and changing it into a gas.

Those energy transfers are easy because of help from Mother Nature. Heat energy flows spontaneously from an area of high concentration (hot body) to an area of low concentration (colder body). The greater the difference, the greater the transfer rate. Of course, this is also what's happening when your cup of tea cools off or when an ice cube melts.

When you dip your fingers into 50°F water, it feels cold because there is a 48.6-degree difference in temperature between the water and your body. Heat energy is moving from your fingers into the water, raising the temperature of the water, if ever so slightly.

All of this is embodied in the Second Law of Thermodynamics, noted below. The temperature *difference* is the key: as long as there is a temperature differential, heat energy will spontaneously flow to the colder object, raising its temperature.

A ground loop containing a flow of water at, say, 33°F (0.6°C) will continuously pick up heat energy from the earth—that is at 50°F (10°C)—and the loop's temperature will increase. This heat energy will be continually passed into the refrigerant loop through a heat exchanger, increasing the temperature of the refrigerant.

> **The Second Law of Thermodynamics**
>
> The second law of thermodynamics deals basically with entropy or the measure of disorder in a system. In brief, it states that heat energy will always transfer spontaneously from a hotter to a colder body and never the reverse.

3) Compression

Next, the refrigerant gas is sucked into a device that increases the pressure (that's the compressor, of course), and the temperature of the refrigerant increases even further, proportionate to the pressure increase. If you've ever used a household pressure cooker, you know that after you add food and water, close it up and turn on the heat that the water starts to boil and turns into a gas. The pressure increases inside the closed cooker and the cooking time is reduced.

This effect is based on what is called the Ideal Gas Equation. It states that in any closed, fixed-volume system, the Pressure (P) is proportional to the Temperature (T). In other words, if P increases (or decreases), then T must increase (or decrease). If P doubles, then T doubles.

The compressor in a GHP applies that Pressure/Temperature law by compressing the gaseous refrigerant, increasing the pressure and increasing the temperature. This becomes another step that further increases the temperature of that refrigerant.

4) A Second Source of Heat Energy

And finally, another source of heat energy is introduced into the refrigerant. The electrical energy that drives the compressor was converted to mechanical energy and then converted into heat energy, as shown in the previous chapter. And that heat energy also increases the temperature of the refrigerant. Now we have two sources of heat energy, both of which are increasing the temperature of the refrigerant. The geothermal heat pump has now achieved super efficiency and we have high enough temperatures—165°F (74°C)—to heat the home.

5) The Final Step

Back to that refrigerant: The refrigerant is in a closed loop, but we have not yet finished that loop. The refrigerant must finally be allowed to reduce its pressure (and therefore the temperature) by expansion. That expansion is the job of the Thermostatic Expansion Valve (TXV). It performs the opposite function of the compressor by allowing the high pressure refrigerant to expand, thus lowering the pressure and (following the P/T relationship) lowering the temperature considerably, sometimes below freezing.

This is exactly what is needed because the lower the temperature, the greater the temperature differential in the ground loop and the more efficient the heat energy transfer becomes. ✪

Chapter 18

The Importance of GHP Performance Standards

If you think that industry standards are only something for design engineers or manufacturers, but not really of interest to the homeowner, think again. Standards are a way of insuring that there is uniformity in how competing products are rated, measured, tested and (especially) compared.

Standards are for consumer protection. Without standards, companies would define product performance differently, test their products differently, and no one could compare one product to another. Mature industries want standards so they can design to meet or exceed them. Once a standard is set, testing methods can be devised for independent organizations to certify compliance of the product to those standards, insuring that ratings are uniform and that there is a fair comparison. Certification is voluntary, but competition is the driving force.

Of the many types of standards that apply to GHP systems—performance, quality, safety and electrical standards—this chapter will concentrate on performance standards. Performance standards in the GHP field relate to efficiency standards. As discussed in earlier chapters, efficiency is the primary reason that GHPs are so much better than any other heating and cooling approach. Performance testing and ratings are done in a lab or factory where they measure the energy output at the heat exchanger. Therefore the results will be lower in your home because there is no way to test for all of the onsite variables, such as ductwork losses that are unique to every installation. The measurements are really conducted at the second stage, the refrigerant loop, because they cannot predict your specific ground loop or your specific heat loop.

Who Sets Standards?

ISO (International Standards Institute), with a Central Secretariat in Geneva, Switzerland, has developed over 18,000 standards and adds some 1,100 new ones every year. It is a non-governmental organization that forms a bridge between the public and private sectors. ISO has a network of national standards institutes in 159 countries, one member per country, plus members from national industry associations. Certification to ISO standards by third-party institutions allows the use of the ISO label.

The ISO 13256 is the international standard for open loop and closed-loop GHP systems. It basically lays out standard testing conditions and requirements.

Energy Star is a US government program, although it has also been adopted by Canada with a few changes to reflect their more severe weather conditions. The US Environmental Protection Agency along with the Department of Energy have set up an Energy Star program that specifically includes GHP systems and their efficiency performance standards. It specifies minimum COP and EER ratings. Energy Star incorporates ISO 13256 and AHRI 870 *(see below)* into the requirements. The reward for agreeing to meet these standards is the right to display the Energy Star label. EPA's intention is to utilize the AHRI Directory of Certified Products to determine which equipment qualifies for Energy Star.

The current TIER 1 Energy Star minimum performance numbers for GHPs are shown below. TIER 2 requirements with increased numbers will be effective on January 1, 2013.

Energy Star Criteria (Tier 1 & 2)
Energy-Efficiency Criteria for Qualified Geothermal Heat Pumps

Product Type	EER		COP	
	Tier 1	Tier 2	Tier 1	Tier 2
Closed Loop Water-to-Air	16.1	17.1	3.5	3.6
Open Loop Water-to-Air	18.2	21.1	3.8	4.1
Closed Loop Water-to-Water	15.1	16.1	3.0	3.1
Open Loop Water-to-Water	19.1	20.1	3.5	3.5
DGX	15.0	15.0	3.5	3.5

* Tier 2 will be effective January 2013

Figure 18-1

AHRI (Air Conditioning, Heating and Refrigeration Institute) is a US trade association representing over 300 manufacturers. It establishes industry standards and offers independent testing and certification to those standards by independent laboratories under AHRI contract. It is a voluntary, non-profit organization representing the major voice of the heating, ventilation, AC, refrigeration, and GHP industry. It established the original US GHP standards that were closely adopted by the ISO as an international standard. They have also established the AHRI 870 standard that is now the DX standard in the United States.

ANSI (American National Standards Institute) is the nonprofit voice of US standards and conformity assessment. It adopts standards and provides accreditation programs to assess conformance with the standards. Headquartered in Washington, DC, ANSI is the official US representative to the International Organization for Standardization (ISO).

CSA (Canadian Standards Association) is a not-for-profit membership-based association serving business, industry, government and consumers in Canada. They adopted the ISO 13256 standard as C13256 for open- and closed-loop systems, while recognizing that their colder climate requires some tailoring to the standards. Those companies who meet the standards are permitted to display the CSA label. In addition, CSA standard C222 covers DX systems. Canada has moved ahead of the United States in establishing a standard for installing GHPs, a very important step. Standard C448 covers the complete installation of open- and closed-loop systems: pipe quality, pipe fusion, foundation frost protection, ducting, drilling, grouting, etc.

Natural Resources Canada, Office of Energy Efficiency (OEE), a Canadian government operation, is a center of excellence for energy conservation, efficiency and alternative fuels. It is mandated to strengthen and expand Canada's commitment to energy efficiency. It offers grants and incentives, workshops for professionals, statistics and analysis, and various publications. It also uses a modified Energy Star system for their colder climate. The Canadian Energy Star label is slightly different from the US label, displaying a bilingual "Energy Star High Efficiency—*Haute Efficacite*." ◉

Are Factory COPs Accurate?

Someday soon GHP homeowners will be able to see their COP or EER efficiencies displayed on their thermostats. It will be like an automotive miles-per-gallon indicator on the dashboard, except it will be COP or EER on the wall. The numbers you'll see, however, will not necessarily be the same as the manufacturer's certified numbers. That is because the efficiency numbers shown on the thermostat would be computed at that moment, as your system runs, and they will vary according to operating conditions.

Neither one of those numbers, however, will actually represent the advertised system numbers either; they will be lower. By international standards and by practical necessity, the manufacturer and the wall meter can only measure the system's efficiency up to a certain point in the system. That point is the final heat exchanger. They cannot measure the heat losses in the third loop in your home.

A good portion of the system efficiency depends on the efficiency of the ductwork (if that is used), the quality of installation, and even the temperature of the water entering the heat pump. So actual efficiencies will be lower than the official ratings.

The factory numbers are still very useful. They show that they meet or exceed required standards and they permit comparisons between different units. The wall meter will help you set your system for optimal efficiency, much as an automobile MPG meter can help you achieve the most miles per gallon from your driving.

The EPA has stated that high-efficiency geothermal systems are on average 48% more efficient than gas furnaces, 75% more efficient than oil furnaces, and 43% more efficient when in the cooling mode.

Part III

The Broader Picture:
A Look Beyond Residential to the Global Market

Chapter 19

The Geothermal Marketplace

If I were an investor (I can only dream), I would look very closely at Canadian and US geothermal manufacturing companies. This is a growth market that will be pushed upward by the new tax incentives in both countries and by homeowners' desires to reduce living expenses.

Looking at the past not only provides some perspective about the future, but it also shows that the geothermal industry has over 60 years of operating experience. The first United States geothermal heat-pump system was installed in the Commonwealth Building in downtown Portland, Oregon, in 1946. This was a successful and highly publicized project that led to a number of installations throughout the country.

But widespread acceptance by architects, engineering firms and builders was very slow. The low cost of fuel and the high cost of the ground loops were barriers. There was a brief growth period during the oil crisis of the 1970s, but it was not until the last few years that commercial, institutional and residential applications have surged ahead as oil costs have increased. Ground-source geothermal heat pumps are now one of the fastest growing applications of renewable energy in the world. In

2008, Canadian heat pump sales increased by over 50%; even more in Ontario. US shipments increased by 40% in 2008 to 121,242 units.

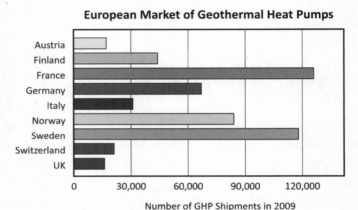

Figure 19-1
Source: *European Heat Pump Association; "Outlook 2010" Report*

However, this technology is currently only a small segment of the overall potential market. Total United States and Canadian new home starts in 2008 totaled over 1.1 million, and that does not include new building starts.

The Oak Ridge National Laboratory 2008 ground source geothermal report (see appendix, *Helpful Links and References*) concluded that although the United States has an installed base of about 600,000 units, the European markets absorb two to three times the number of GHP units per year than the United States. It is a worldwide growth industry, with rates in Europe and parts of Asia and Canada exceeding those in the United States. Sweden and France are the largest heat pump markets in Europe. In 2009, 524,000 units were sold in Europe; quite a jump from 92,000 in 2004.

19 — The Geothermal Marketplace

The geothermal market is a part of the Heating, Ventilation and Air Conditioning industry, commonly referred to in the yellow pages as HVAC. The 2007 total HVAC market in the U.S. was $14.3 billion.

Ground-source GHPs are a $2.5 billion dollar industry in the US that is growing by 30–40% per year as oil prices have surged. In the United States 36,439 GHP units were shipped in 2003, 83,396 in 2007, and 115,442 in 2009 for residential, commercial and institutional applications. So you can see that current GHP shipments are a small percentage of the total potential market. The industry's goal is to capture 30% of that heating and air conditioning market by 2030.

The current growth is producing growing pains, especially in terms of having enough qualified installers, drillers, and even sufficient materials, such as high-density polyethylene. But growth in the geothermal market involves a large number of small businesses, which translates to additional benefits for the local economies. It is a win-win situation for everyone. ❂

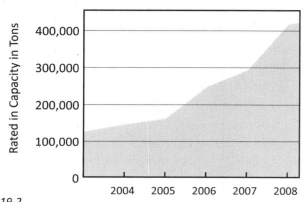

Figure 19-2
Source: *Energy Information Administration, "Annual GHP Manufacturers Survey" Report*

Chapter 20

Commercial / Institutional Geothermal Systems

While this book is primarily directed toward residential applications, there is a growing and parallel effort underway for applying ground-based geothermal technology to a wide variety of government, institutional and commercial buildings, and even non-building applications. What made this so interesting to me was the amazing diversity of applications; applications that I had never considered.

Since projects of this type are usually larger in terms of energy, they have a larger impact on energy use. GHP systems provide greater savings for commercial buildings with their large heating and cooling load. The payback period is usually four to eight years, but it could be as short as two to three years. The first United States geothermal heat pump system installed in 1946 was a commercial installation.

Today's commercial GHPs are considered a cost effective, energy efficient and environmentally friendly way of heating and cooling buildings. They are appropriate for new construction as well as retrofits of older buildings. Their flexible design requirements make them a good choice for schools, churches, high-rises, government buildings, apartments and restaurants.

Commercial installations also have significant advantages with geothermal:
- Different parts of the building can be simultaneously heated and cooled. For example, the sunny side may need cooling while those on the shady side need a bit of heat.
- The quietness of operation is a benefit to the building's occupants; they won't even know when the system is running.
- Multiple zones allow individual room control.
- Less mechanical space is required (50–80% less!).
- No outside equipment to hide (and no rooftop units), which also eliminates vandalism.
- All-electric system eliminates multiple utility services.
- No boiler and chiller maintenance.
- Lower life-cycle costs.
- Lower peak demand and lower operating costs.
- Big savings on energy consumption, from 25–50%.
- Water for consumption can be heated with waste heat in the summer at no cost, plus reducing costs in winter.
- Rebates and tax incentives available.

Canada, with cooler weather and lower average temperatures which results in higher heating costs, has been adept at finding innovative applications for ground-source geothermal technology, including fish hatcheries, tree nurseries and curling rinks.

Agricultural and horticultural businesses also use geothermal energy in a variety of applications. For example, dairy farms use GHPs for milk pasteurization and chilling as well as hot washing of equipment. Greenhouses in several western states use direct thermal wells for heating, while New England and Canadian greenhouses are successfully applying heat pump technology for this application. Large-acreage tree nurseries are typical examples where fuel savings of over $10,000 per year are being obtained.

Here are just a few other examples that show the wide range of geothermal applications:

Smart Bridges—Research is underway in Oregon and Oklahoma to use GHPs to circulate heated fluid through tubes embedded in bridge decks to reduce icing, which in turn, will eliminate salt usage and the resulting corrosion. These 'smart' bridges are tied into a network of weather stations to predict icing conditions.

Gas Stations—One major gas station/convenience store chain is retrofitting one store per week with geothermal systems. These systems absorb waste heat from freezers and icemakers and even provide hot water for a car wash area.

Government Buildings—At the west entrance of the Colorado State Capitol in Denver, the fifteenth step has a plaque engraved "One Mile Above Sea Level." In 2010, deep beneath those steps, two 900-foot (274-m) holes were drilled for an open-loop, $6 million geothermal heating and cooling installation that will provide an estimated savings in utility bills of $95,000 per year for the state. The supply well taps into the 55°F (13°C) ground water heat of the Arapahoe aquifer; the injection well receives the water after the heat has been extracted.

Hotels—One of the largest GHP installations in the United States is the Galt House East Hotel in Louisville, Kentucky. Heat and cooling are provided for 600 hotel rooms, 100 apartments and almost a million square feet of office space. The hotel saves $25,000 per month compared to a similar non-GHP building adjacent to it.

Commercial Profile: Baxter Building

The Baxter Building is a 14,000 square-foot (1,300 sq m), mixed-use office building in Poughkeepsie, New York, that proves DX GHP systems are equally applicable to commercial buildings. It uses a DX geothermal heating and cooling system because the cost of energy for a traditional HVAC—which would've required over 4,000 gallons (15,141 liters) of oil each year for heating—made the property owners and developers look for alternatives.

The builder, R.L. Baxter, received geothermal bids from two companies. One bid specified a closed-loop system using deep wells with over 8,000 feet (2,438 m) of cased well drilling. The other bid from Total Green Geothermal of Monroe, New York, was accepted. They offered a DX approach with higher conductivity, more efficient copper piping and refrigerant in the ground loop. Six DX fields were placed in the parking lot and covered with environmentally friendly asphalt that allows drainage through the surface (this moisture actually allows the fields to perform better by improving heat conductivity). Fifty-six holes, each 70 feet (21 m) deep, were drilled at 30-degree angles. Advanced Geothermal Technology from Reading, Pennsylvania supplied four 5-ton and two 4-ton DX units.

Using green technology not only saves money, but assists in the marketing of such buildings.

Schools & Colleges—Recently the US Department of Energy reported that over 500 schools in the United States have installed geothermal heating and cooling systems. The ability to cool school buildings permits year-round occupancy, if needed. The Comanche Elementary School in Comanche, Oklahoma has an interesting GHP installation because it uses the city water supply as its heat source to get very energy-efficient heating and cooling. This idea appears to open new city applications where trenching is not an option.

The Australian Outreach College in Brisbane, Australia has installed a direct exchange (DX) geothermal system to cool classrooms and an integrated- technology building that has high heat loads from people and computer equipment.

Bard College in New York State has built a set of dormitories heated and cooled with a GHP using vertical boreholes.

Hospitals—The new Sherman Hospital Medical Center near Chicago uses a 15-acre (6-hectare) lake as a heat source for heat and cooling, claiming a savings of over $1 million a year in energy costs. The hospital received a $400,000 grant from the State of Illinois and $956,000 from DOE.

Airports—As part of a major upgrade, the Juneau, Alaska, airport has installed in 2009 a ground-source geothermal heat pump system that is expected to reduce energy costs by $85,000 per year. That same year, the Nantucket Memorial Airport completed a similar system.

Churches—In the US, Canada and throughout Europe you can find stories of both large and small, very old and very new churches being heated and cooled using heat pump technology. An interesting twist occasionally resulted. In retrofit examples where insulation was inadequate, the general results show that the annual heating costs were not lowered after a GHP installation. What was gained, however, was cooling at no extra cost, providing a higher level of comfort and increased attendance and contributions in the summer. As temperatures generally increase around the world, this will become even more important.

Trinity Church, Boston, MA

In the late 1980s, my wife and I decided to experience city life and what better city to choose than Boston? We bought a newly renovated 1850-era South End condominium where we had the first floor, lower level and a rear garden. It was within walking distance of Martha's art studio in an ancient former distillery building. It was also close to the Charles River, the Esplanade, the Public Gardens and the golden-domed State House for walking and bike riding.

And then there was Copley Square where, when standing at the Boston Public Library, you could see across the plaza to one of the most beautiful churches in the United States, Trinity Church—a Boston historic landmark built in 1877. Although not a member, I often stopped in to see the sunlight through the stained glass windows, the sculpture by Saint-Gaudens, the murals, or just sit and enjoy the quiet atmosphere.

That Back Bay area consisted of filled, marshy wetlands in the 19th century so that the architect, H.H. Richardson, had to build the massive 9,500-ton church on four large granite pyramid-shaped piers that sat atop thousands of wood pilings. After 125 years, a great number of repairs were needed. And, because of a changing water table, many of the exposed pilings were deteriorating, threatening the integrity of the building.

After years of planning, a complicated and massive restoration/expansion project was initiated in 2001. The design and engineering teams struggled with where to put the new mechanical systems for the expansion. The church's steep roofs and spires prohibited a typical roof-mounted cooling system. Instead, they chose a geothermal energy system. Six wells, each 1,500 feet (457 m) deep, were drilled within feet of the structure. Thirteen heat pumps totalling 130 tons of capacity were installed. Estimated savings are an impressive $67,000 each year.

Australian Government Science Building—The Geoscience Australia building in Symonston, Australia, is the largest ground-source geothermal installation in the southern hemisphere (and maybe the world). It uses 210 heat pumps and 352 boreholes, each 320 feet (98 m) deep. It has been operating for 10 years.

Canadian Steel Plant Sludge Drying—To dry sludge for more economic transfer to a landfill, a closed-loop geothermal heat pump is used to heat and dry air that is passed through the sludge to pick up moisture which is then extracted by a condenser.

Seawater-Source Heat Pump—A 28,000 square-foot (2,600 sq m) research center in Norway has installed a geothermal heat pump that uses seawater as the heat source and ammonia as the refrigerant.

Tree Nursery—A New Brunswick, Canada nursery uses four 35-ton GHP units for greenhouse heating.

Fish Farm—A fish hatchery in Canada employs a GHP to raise the incoming fresh water temperature for optimal hatching and growth.

Civic Center—The Port Hawkesbury Civic Center in Nova Scotia, Canada, is a multi-purpose building of 22,660 square meters (244,000 sq ft) that uses a geothermal heat pump. An Ice Kube geothermal chiller cools the ice rink and uses the heat for radiant floor heat, melting of sidewalk snow, and melting of ice shavings from the rink. Forced air heat pumps also provide heating and cooling. ✹

Chapter 21

Future Trends

As with most technologies, competition forces companies to invent improvements to gain a market edge. Many interesting new trends are in the works.

Increasing GHP Efficiency

The unusually large efficiency of a GHP system is what makes it such a powerful candidate for home or building heating/cooling. But the GHP level of efficiency has not yet peaked. In the coming years, I expect it will increase past 600%, perhaps even 700%, as engineers find ways to extract more heat energy out of the ground and components are improved to do the same amount of work using less electric power.

Direct Financing

Contractors and manufacturers are starting to offer 100% financing so that the system owner is not tasked to find bank loans.

Geothermal Leasing

Electric utilities in Minnesota and Colorado are now installing, paying for and leasing back the ground loops for customers'

GHP systems for a fixed monthly tariff. This is smart, long-term thinking and a huge step in making GHPs more affordable. The business logic is that the ground loops can be considered a utility investment much like power poles and wiring. Indeed, with many people moving from existing homes in five years or less, the concept of leasing the entire GHP package is actually under consideration (or at least up to the ductwork, radiant floor or hot water heat delivery subsystem). This would eliminate the burden of financing and would probably be a package that includes all service, repairs and maintenance.

On-Board Telemetry

On-board telemetry will allow the owner to monitor the actual energy consumption on a daily or monthly basis to verify system benefits and control costs. It will be like the miles-per-gallon meter in your car. This concept can be expanded to calculate annual greenhouse emissions to aggregate credits for the cap-and-trade market where and if it goes into effect.

Variable-Speed Compressors

The first compressors had one speed no matter what size of load. Then two-stage units permitted the compressor to coast at a lower level to meet lower needs, saving electricity, while the second stage was available for higher loads. Expect to see infinitely variable stages that will automatically adjust to the exact level needed.

Solid-State Compressors

It all started with a piston-type mechanical compressor with over 20 moving parts requiring oil for lubrication, then converted to a higher-efficiency scroll compressor with only six moving parts and then to a digitally controlled scroll compressor. What is inevitable in the long term is a solid state device with no moving parts and no oil required. Research is also underway in magnetic pulse compressors, electro-chemical approaches, superconducting cryogenic devices and several other concepts. It will first be applied to refrigerators and then to GHPs, but it will not be here next week.

Better Building Efficiency

The growth of geothermal and solar will increase the homeowner's interest in building efficiency and load reduction. There is an increasing trend for contractors to look at the total picture: house design, solar, geothermal and wind installations.

A Zero Energy Home (ZEH) is a general term applied to a building that has net zero energy consumption and zero carbon emissions annually. A ZEH combines high levels of energy efficiency with state-of-the-art renewable energy systems to annually return as much energy to the utility as it takes from the utility, resulting in a net-zero energy consumption for the home. A geothermal heat pump fits perfectly into Net Zero and green building goals.

A "green building" is commonly defined as using architectural resources more efficiently and to reduce a building's impact on the environment. The LEED (Leadership in Energy and Environmental Design) system is an internationally recognized building certification program developed by the US Green Building Council (*www.usgbc.org*). It provides check lists as measurement tools to achieve green building goals that can be used for third party verification. It applies to both commercial and residential buildings.

Net Zero Homes and green buildings can deliver one important goal: to completely or very significantly reduce energy consumption and greenhouse emissions for the life of the building. This involves the following disciplines:
- Don't build more space than you need
- Build the most efficient building you can afford (the best windows and insulation, optimum orientation for passive solar heating, etc.)
- Add solar electricity and solar water heating to the mix
- Reduce energy demand

To learn more, visit the National Home Builders Association's Green Building Program: *www.nahbgreen.org*

Shortage of Qualified Installers

As incentives kick in and public awareness increases, the number of GHP requests will accelerate. A lack of qualified contractor/installers may become a problem. Here is where jobs will be looking for people. Accreditation courses are necessary but cannot make up for the lack of hands-on field experience and apprenticeship.

Official Installation Standards

Standards for GHP hardware compliance are in use everywhere, including efficiency standards, but there are no officially approved national AHRI *installation* standards as of 2010. Until this is addressed, the International Ground Source Heat Pump Association (IGSHPA) provides an installation standard that is used for accreditation training of installers in the United States. Canada has adopted standard C448 as their official installation standard. ✪

Chapter 22

Decision Time

Now that you have read this book, you have an edge; an edge over those homeowners who are unable to take advantage of GHP technology because of lack of information. You now understand that you can save serious money every year by not buying fuel oil, natural gas or propane; you can reduce pollution by not burning those fossil fuels in your home; and at the same time, you can reduce your maintenance costs.

You are now in a position to join the thousands of homeowners all across North America who are convinced that they have made a very smart decision—installing a system that is amazingly reliable during the coldest or the hottest weather; one that is proving to be the most efficient and the most comfortable way to heat and cool their homes.

You can now appreciate the fact that current geothermal heat-pump tax credits will not last forever and, therefore, now is the best time to take advantage of these incentives that make GHPs affordable for any home, new or retrofit, in any location and climate.

And to top it off, you also know where to start by finding an accredited, experienced heat pump contractor.

My prime motivation for writing this book was to increase public awareness of this technology. And in doing so, I may have actually provided more than you really need to know about geothermal heat pumps, but that was deliberate. You deserve the bigger picture.

In making your decision to go geothermal, I urge you to think long-term. Your first concern is probably cost—initial costs, annual costs, and payback time. I read a *New York Times* article recently about a survey of corporate CEOs. They were asked what decision they would make on an important program if it meant losing short-term profits and but gaining long-term profits that helped insure the continuity of the company. Eighty percent decided that short-term profits were more important! Their personal rewards were based on short-term results. Sadly that sort of thinking contributed to the financial crisis of 2008–09. We all need to look beyond next week.

I will restate this important point: **A geothermal heat pump is superior to any other method of home heating** in terms of combustion (it does not combust fossil fuels), renewability (uses free renewable energy from the earth), and cost of operation (big savings can be had).

Installing a GHP is almost always in the homeowner's best interest and definitely in our country's interest.

Good luck. Good heating. Go Geothermal!

My Ultimate Geothermal Package

This book has explored the subject of ground-based geothermal systems from many directions. Now, it is time to tie things together and end this story with my version of the "ultimate," or ideal system configuration—the grand geothermal system that does everything—well, except eliminate my junk mail. It is not, however, the lowest cost solution.

First, I needed a set of criteria. How does one define the ultimate solution?

1) It will not only require no flame to operate, but also will be operated on net-zero electrical power. That is, over a year it has zero electrical cost.
2) It will have the highest efficiency, or a COP = 5.8.
3) It will be very reliable and require very low maintenance.
4) It will provide full domestic hot water—no electric or gas-fired hot water tank.
5) It will have a reasonable payback period.
6) It will eliminate the anxiety of losing power (and therefore heat) in a heavy snowstorm.

So, what would my ultimate system consist of? A **DX (Direct Exchange) system** with a digital scroll compressor and full geothermal tankless domestic hot water meets Items 2, 3, and 4. It also contributes heavily to Item 1 with its low electrical operating cost and to Item 5 with fewer heat exchange loops.

A **solar electric installation** will meet Item 1 by reversing the electric meter. A small south-facing photovoltaic array may fall behind during cloudy winter days, but will come out even over the year.

A **propane-operated standby emergency power generator** of sufficient size will meet Item 6.

Epilogue

The Disappearing Stars

There was no moon. There were no clouds. But the clear night sky was filled with billions of glistening diamonds. I kept looking up, pondering the decisions we had made, wondering how it would all work out. This was our last night in the schoolhouse, the movers coming in the morning, the new owners anxious to move in. We were moving to an apartment for six months, close to the building site. The architectural plans were complete, the builder was starting work, and a 4-ton capacity geothermal system was on order.

I have found myself looking up at the stars for answers many times in the past when changes were coming or uncertainty was nagging. As a small boy about to enter a new school in a new city, as a college student temporarily stranded and hitchhiking on a Connecticut highway at night, as a soldier in basic training looking up at the Texas night sky wondering if I would survive the war, as a junior officer in war-torn Germany thinking about my future, and many years later in the Loire Valley, savoring the French culture, language and food while trying to figure out how we could make that deserted schoolhouse back in the in New York countryside into our home. And now, I was wondering if I had made the right heating decision for the new house.

For centuries, the stars have been helping humans navigate and find direction. For me, gazing at the stars seems to help me clarify my thinking and find my own direction.

Sophisticated techniques and advanced mathematics have been used to make measurements of those stars that have revealed much of the past and some of the future of the universe. In the long run, scientists now believe that most of those stars will start to blink out, disappear from our view. There is a story here that starts with Albert Einstein.

In 1916 Albert Einstein published his General Theory of Relativity. But he was unable to believe his own equations because they showed that, due to the force of gravity, the universe could not be static. And yet the scientific community knew for a fact that there was a "Steady State Universe"—always had been, always would be. So he inserted a fudge factor into his equations, a small amount of force that mathematically stabilized the universe against expansion or contraction. He called this the Cosmological Constant. Now his theory agreed with the perceived reality of the universe and scientific peace reigned.

That is until the late 1920s when Edwin Hubble formulated Hubble's Law that led to a widespread acceptance that the universe was actually expanding, that the galaxies were moving away from each other. From Einstein's view it was "the biggest blunder of my life." As he removed his Cosmological Constant from his equations, he metaphorically slapped himself on the forehead because he could have predicted the expansion. But of more importance, there was then a complete change—an upheaval—in cosmological thinking because an expansion could be mathematically worked backwards in contraction. That, of course, led to the Big Bang theory when all the matter in the universe was concentrated at a single point of infinite density as time began 13.7 billion years ago.

But the story was not over. In the past 10 years, astronomers have discovered that the universe is not only expanding but that the expansion is speeding up. They believe that this accelerated expansion will eventually pull galaxies apart faster than the speed of light. That means that they will drop from view because their light could never reach us. We will see no stars except from our own galaxy. It also requires that Cosmological Constants be reinserted back into the equations.

In the world of cosmology, physicists have gained immense amounts of information to learn how the universe works. But they still have no idea *why* the universe works or where it came from. If there was a Big Bang, what happened before that?

Now, a novel approach called Null Theory by Terence Witt is annoying the establishment. Witt states that the universe is infinite and eternal, that the universe is not expanding, and that the Big Bang never happened. He also states that current theory violates the Law of Conservation of Energy while his theory is in consonance. As he puts it, current theories use the red-shift of light from distant stars to measure the expansion and acceleration of the universe and these theories require the invention of dark matter and dark energy (more cosmological features that have never been observed or measured) to make the equations agree with the observations. Witt claims that the red-shift observations are not a measure of expansion, but are from gravitational effects on the light. In his opinion, this makes the "whole expansion concept a fantasy because the universe is infinite and will not expand, contract or grow old and die."

I am not qualified to judge his explanations as right or wrong, but this subject makes for an intriguing discussion. It is so elemental and thought-provoking. I can barely wait for the next chapter.

Although not obvious, there is a bond between the micro-universe of the geothermal system in my basement and the macro-galactic universe out there in space. Both are subject to the same universal and empirical laws of nature such as the First Law of Thermodynamics, also called the Law of Conservation of Energy, which states that the total energy in an isolated system cannot be created or destroyed, but it can be converted from one form of energy to another form. I used that law to demonstrate *why* a GHP works by changing electrical energy into mechanical energy and that mechanical energy into heat energy. I carefully separated the question of *how it works* from *why it works* to better demonstrate this. In the cosmological field, the laws of nature prevail but an understanding of the How and the Why is still in flux.

While all of this goes on, I am looking forward to the next clear night to again ponder the future of the heavens—and my own. ✪

Appendix A

Frequently Asked Questions

How It Works

Can one geothermal system provide both heating and cooling for my home?
Yes. One of the advantages of a GHP is that instead of two separate systems, one push of a button on the thermostat switches from heating to cooling.

Can a GHP be installed in new homes as well as older homes?
Yes. GHP systems are being installed every day in new construction as well as in older homes.

What are the alternatives for the basic configurations of earth loops?
They fall into two categories: open loop and closed loop. **Open loops** take water from a pond or well, extract the heat and then return it to the same or another location. New water is constantly being used. **Horizontal closed loops** circulate water in plastic pipes placed in trenches 5–6 feet (1.5–1.8 m) underground and then return it to the GHP, extract the heat and then recycle the water back into the ground in a continuous operation using the same water. **Vertical closed loops** send the water into vertical bore holes that may be as deep as 70 feet (21 m) to perhaps 600 feet (182 m) and then return it to the GHP. **DX or Direct Exchange** systems use copper piping instead of plastic and send the refrigerant through the ground and back. Instead of having a separate ground loop and refrigerant loop, they are combined into a single loop. See chapters 4 and 5 for details.

If I use my well in an open loop system, what capacity or flow rate must I have?
Your manufacturer will specify the flow rate needed for your system, but the rule of thumb is 2–3 gpm (7.6–11.3 lpm) per ton of GHP capacity. So, a 4-ton unit needs at least 8 gpm (30 lpm). Now, your domestic needs must be added in. Many people find that 2 gpm (7.5 lpm) is adequate.

What are the advantages and disadvantages of these various configurations?
Open loops are generally less expensive because there is less trenching. Horizontal closed loops require more available land space. Vertical bore holes require less space, but are more expensive to drill. DX systems are usually more efficient and require less ground space since the holes are

drilled at angles from a central point. Chapter 7 has additional information.

How far apart are the trenches or vertical boreholes placed?
Ideally, vertical bore holes should be at least 15 feet (4.5 m) apart, preferably 20 feet (6 m) apart. The centers of each horizontal trench should be about 10 feet (3 m) apart.

Can I install my own geothermal system?
Yes, if you are a certified, trained and experienced installer. Otherwise, forget it. Read chapter 15 for why this is important.

How much space should I allow for the interior equipment?
The floor footprint of the heat pump is about the same as a refrigerator, but ductwork and piping can double that.

How can a GHP heat my domestic water?
A desuperheater is an optional add-on with a separate heat exchanger that can provide excess heat to your hot water tank. Chapter 10 describes this plus other methods of obtaining hot water.

Is a geothermal heat pump noisy?
No, a GHP is very quiet, especially compared to fossil fuel burners with their abrupt, noisy start up while sucking in huge amounts of air.

Are GHPs safe?
Well, they are a little hard to steal. Sorry, could not resist that. The lack of any combustion means no flames, no noxious gases. They are as safe as your refrigerator.

Can a GHP be connected to an existing radiant system?
Absolutely, and also to an existing hot water system, or a hot air system. This can save considerable expense. Radiant or hot water will not permit heat pump air conditioning, of course. For hot air, you should check with your contractor. Many fossil fuel systems have a high velocity air flow that use smaller-sized ductwork. A GHP saves money by using a slower, more comfortable rate of flow, but requires a larger sized duct.

Will an underground horizontal loop affect my landscaping and lawn?
With horizontal trenches, yes, but only for a week or so. The grounds will soon be back to normal. It is also possible to bore underground horizontally if your contractor has the proper equipment. This reduces the impacted area considerably. See chapter 4.

What level of maintenance is required?
Very little maintenance is needed. There is no outside equipment, as

with an air conditioner; no annual tune-ups or nozzle replacement, as with an oil burner; and no chimney to clean. Air filters need replacement as with any air delivery system. Water filters need cleaning if you have an open-loop source.

Do I need a programmable thermostat with my GHP?
Not really, but it will not come without one. You gain a lot of flexibility, plus the installer needs it to set a number of important parameters. A GHP thermostat does a lot more than just set temperatures. For example, with the push of a button (or automatic programming) it switches from heating to cooling mode. For maximum efficiency, experts recommend that you set the GHP thermostat and never change it. Each zone will come with its own thermostat. You'll find the programming capability really quite handy.

Do systems in cold climates need additional heat sources?
While properly installed GHP systems will provide all the heat or cooling needed, all systems require an emergency backup system. For a possible compressor failure, there is a backup electric subsystem built-in. For a power failure, the GHP is in the same unhappy position as any standard system. A backup emergency device is very comforting.

The Nuts & Bolts of the Technology

How do you transfer heat energy from the earth?
The earth has the ability to absorb and store heat energy from the sun. If we want to use that stored earth energy, we can extract heat through a liquid medium that is colder than the earth temperature and then transfer that heat energy to the heat pump. Heat will always flow spontaneously from a warmer medium to a cooler medium.

Will a GHP lower the earth temperature if we keep pulling out heat energy?
No, the earth absorbs 47% of the sun's energy that reaches the earth. This is 500 times the total energy needs of the planet. Locally, the ground temperature can be modified temporarily, merely reducing the efficiency a bit. But during summer cooling, the heat energy is returned to the earth.

I put my hand in 50°F water and it was very cold! How can that possibly heat my home?
Any material at 50°F contains a large amount of heat energy. It feels cold because your hand temperature is near 98.6°F. Heat energy is being transferred from your hand to the water, warming the water. And that is because Mother Nature insures that heat

energy spontaneously flows from the warmer to the cooler object. But assume for a minute that you remove your hand and insert a pipe containing a liquid at 40°F. The heat in that 50-degree water would be transferred to the cooler refrigerant.

But I still do not see how that 50°F water ends up providing 72°F room temperature if I'm not burning anything.
Actually, that 50-degree water ends up providing much higher temperatures than 72 degrees. To make a room feel comfortable, we must provide hot air that is higher than skin temperature, or about 100–105°F. To reach that temperature, we must pass the air over a heat exchanger that is about 165°F to compensate for the heat it loses as it passes through many feet of cooler metal ductwork. By using a compressor and a series of valves and controls, your geothermal energy system works its magic to give us very high temperatures for home heating. See chapters 16 and 17 for a detailed explanation of what goes on inside the GHP.

How does a geothermal heat pump cool the home?
By pushing a button on the thermostat, you can immediately switch from Heat to Cool. This activates a reversing valve that changes the flow of heat energy from your home into the ground.

Why does the geothermal ground loop need antifreeze when it is below the frost line?
The water entering the ground loop often drops below 30°F (1°C) after transferring its heat. The lower the temperature, the better the heat transfer from the ground. Heat transfer requires a temperature differential; the greater the difference, the more efficient the transfer.

Are we wasting water with an open-loop system?
No, we are just returning it back to the aquifer where it came from.

If I use a water softener, should the open-loop geothermal water also be softened?
No, do not condition water for the GHP or for water used for outdoor purposes.

How do I determine how efficient my GHP unit is?
Look for the Energy Star label or go to your manufacturer's web site. The Coefficient of Performance (COP) indicates heating efficiency. Your unit should have a COP of 3.3 to 3.6 or greater. Mine is 4.1 and that means it is 410% efficient. Chapter 18 provides much more information including definitions about GHP efficiencies.

Savings / Costs / Warranties

How much does a GHP cost?
Buying a heating/cooling GHP system is not like buying a refrigerator off the shelf. Each installation is unique and costing requires a qualified contractor. Chapter 12 provides more details. For example, a 4-ton unit plus a water heater, two zone controls, electric backup heater, desuperheater, labor etc. cost about $14,000 in 2009. Extra costs include piping, ductwork and ground trenching or drilling. Visit manufacturers' websites for their online savings/cost calculators.

Will a GHP reduce my living expenses?
A GHP will significantly cut your annual heating/cooling costs every year. In addition, the costs to maintain the system will be lower.

What financial incentives are provided?
Substantial US federal incentives provide a tax credit of 30% of the cost for GHP systems, but these tax credits are set to expire in 2016, unless the government votes for an extension. Your state or power company incentives can to found online at the Database of State Incentives for Renewable Energy (*www.dsireusa.org*). Canadian incentives are changing in 2010 (*www.oee.nra.gc.ca*).

More information can be found in chapter 13.

What is a typical payback period for a GHP system?
Payback is defined as the time in years that it takes to recover the added costs of a GHP compared to a standard heating plus cooling system. It can range from zero to 5 years or more. Chapter 14 explains more.

Does a GHP increase the value of my home?
Yes, it does; the EPA has stated that it increases $20 for every dollar of energy saved. If you save $2,000 per year by not buying fuel oil or natural gas, the increase in your home's value will be $40,000. Whether that is real or not depends on the housing market and the buyer, of course. Please refer to chapter 8 for a more rigorous discussion of this.

How long will a geothermal system last?
While a standard oil burner has a 12–15 year life, a geothermal heat pump can last 20–25 years. An oil burner has large bursts of energy followed by sudden, noisy air intakes and chimney exhausts that stress the system. GHPs have smaller changes, fewer working parts, and do not combust anything. Refrigerators using the same heat pump technology can last 30 years or more.

Do geothermal systems have warranties?
GHP manufacturers offer equipment warranties, varying from 1 to 5 years, or more. This is one criteria for judging/selecting a manufacturer. Appendix C lists manufacturers in the US and Canada along with their warranties. In addition, you should get an installation warranty from your local installer.

Environmental Benefits

How does geothermal energy benefit our environment?
This is important and should be well understood. All other methods of heat/cooling your home use up energy sources that are not renewable or produce pollutants into the atmosphere. A GHP emits no pollutants on-site and uses energy in the earth that is continually renewed by nature.

Appendix B

Helpful Links & Resources

US & Canadian Government Agencies

www.dsireusa.org
Database of US State Incentives for Renewables & Efficiency is a comprehensive source of information on state, local, utility and federal incentives and policies that promote renewable energy and energy efficiency.

www.eere.energy.gov/geothermal
US Department of Energy, Energy Efficiency and Renewable Energy website includes geothermal information and education for consumers, DOE blog to ask questions, database on funded programs, and incentives.

www.energy.gov
US Department of Energy

www.energystar.gov
US EPA listing of all companies (including heat pump manufacturers) who are Energy Star Partners, plus much more.

http://oee.nrcan.gc.ca
Office of Energy Efficiency (OEE) of Canada manages the ecoEnergy Efficiency Initiative, Center of Excellence for energy conservation and efficiency. Lists Canadian federal, provincial, territorial and municipal incentive programs.

Geothermal Heat Pump Organizations and Websites

www.ahri.org
American Air Conditioning, Heating and Refrigeration Institute(AHRI) certifies heating, AC and geothermal efficiencies; has a database of certified companies and products.

http://cgec.ucdavis.edu
California Geothermal Energy Collaborative addresses large-scale geothermal power as well as GHPs.

www.csa.ca
The Canadian Standards Association (CSA) is a non-profit membership based association that develops standards, tests and certifies products for Canada and US markets and trains and accredits personnel.

www.earthcomfort.com
Michigan Geothermal Energy Association

www.geoexchange.ca
Canadian GeoExchange Coalition (CGC) promotes public awareness, training, accreditation and certification to CSA standards.

www.geoexchange.org
Geothermal Exchange Organization is a non-profit, trade association sponsored by the Geothermal Heat Pump Consortium for the geothermal heat pump industry. They educate consumers and professionals with online certification programs/workshops, provide a comprehensive GeoExchange Directory of contractors in the US and Canada, and conduct GeoExchange Forum online to discuss geothermal issues.

www.geoexchangebc.ca
GeoExchange BC, a non-profit industry driven organization located in British Columbia, provides education, training and certification in association with the CGC.

http://geoheat.oit.edu
The Geo-Heat Center at the Oregon Institute of Technology Center in Klamath Falls conducts certification and training courses in geothermal and solar PV technology and provides information, case studies, workshops, technical papers, etc.

http://geothermal.marin.org
Geothermal Education Office

www.greenbuildingtalk.com
Find homeowner questions, reports on experiences, info on heat pumps, comments on companies, etc.

www.gogeonow.org
Colorado Geo Energy Heat Pump Association

Appendix B — Helpful Links & Resources

www.heatpumpcentre.org
International Energy Agency's information service to accelerate use of heat pumps, conferences, newsletters, workshops, etc.; based in Sweden.

www.heatspring.com
The HeatSpring Learning Institute in Cambridge, Massachusetts, is an education company providing clean energy training in geothermal heat pump and solar systems.

www.hvac-for-beginners.com/geothermal-ratings
A site that ranks GHP companies, mostly by warranty.

www.igshpa.okstate.edu
The International Ground Source Heat Pump Association, located on the campus of Oklahoma State University, is a non-profit organization established to advance GHP technology through technical conferences, a database of state incentives, geothermal research, and installation training.

www.iowageothermal.org
Iowa Geothermal Association

www.minnesotageothermalheatpumpassociation.com
Minnesota Geothermal Heat Pump Association

www.nahbgreen.org
National Green Building Program by the National Home Builders Association

www.renewables.ca
Online newsletter on renewable energy activity in Canada.

www.wisgeo.org
Wisconsin Geothermal Association

Other Reference Material

Geothermal Heat Pumps: A Guide for Planning & Installing by Karl Oschner, *www.earthscan.co.uk*

Got Sun? Go Solar 2nd Edition by Rex Ewing and Doug Pratt, *www.PixyJackPress.com*

Ground Source Heat Pump Analysis published by RETScreen International Clean Energy Support Center of Canada (*www.retscreen.net*); an electronic textbook for professionals and university students.

An Information Survival Kit For the Prospective Geothermal Heat Pump Owner by Kevin Rafferty, PE, distributed by HeatSpring Learning Institute; *www.heatspring.com*

Oak Ridge National Laboratory Report, 2008. Geothermal (Ground Source) Heat Pumps; *www.zebralliance.com/docs/geothermal_report_12-08.pdf*

Appendix C

Residential GHP Manufacturers
as of December 2010

The following is a list of US and Canadian manufacturers of residential ground-source geothermal heat pumps. In addition to providing their contact information, this list also gives you a sense of the differences between the companies. I've made no attempt to judge their quality or performance. For up-to-date information and to find new companies, please check the internet.

All of these companies offer scroll compressors and most have dealer locators on their websites. Whether they manufacturer their own heat pumps in-house or use private-label heat pumps in their GHP systems, they all provide contractor support and product warranty and service. So what are the differences?

1. Residential Warranties: There is quite a variation in warranties: from 1 to 12 years, even 20 years in one case. Labor is guaranteed by some companies, not by others. Most companies list warranties on their web sites or are pleased to provide information.

As a baseline, you could consider that the US EPA Energy Star Partner program not only requires each company to meet specific minimum performance criteria, but in the past, has also required a warranty of 2 years on parts and labor and 5 years on the refrigerant circuit before they can use the Energy Star label. This is no longer in effect, but it is a good reference criteria.

2. Training and installer accreditation: GHP manufacturers (with a few exceptions who offer turn-key installations) do not sell to the public. They rely on local contractors and installers who must be trained and accredited. When a manufacturer provides accreditation courses, it adds more qualified installers into the field who might well be more apt to use that manufacturer's product. It is a smart strategy. Of course, there are other sources for accreditation.

3. Efficiency Certification to Standards: All companies must (for competitive reasons) have some form of product certification to international, US or Canadian standards, a few to all. Not all are Energy Star compliant. (*Chapter 18 – The Importance of Standards* explains the various certification standards.)

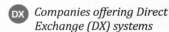 *Companies offering Direct Exchange (DX) systems*

US Geothermal Residential Heat Pump Manufacturers

Bard Manufacturing Co.
1914 Randolf Drive
Bryan, OH 43506
www.bardhvac.com
(419) 636-1194
> Climate control, air conditioners, oil furnaces and heat pumps
> *Certification:* ISO 13256, Energy Star
> *Warranty:* 5 years on parts including compressor.

Bryant Heating & Cooling Systems
7310 W. Morris St.
Indianapolis, IN 46231
www.bryant.com
(800) 428-4326
> *Certification:* Energy Star
> *Warranty:* 10 years on compressor and major parts; no labor; 5 years on other parts; extended warranty available.

Carrier Corporation
One Carrier Place
Farmington, CT 06032
www.residential.carrier.com
(800) CARRIER
> Add-on water heating component.
> *Certification:* Energy Star
> *Warranty:* 10 years on compressor; 5 years on other parts.

Climate Master
7300 SW 44th St.
Oklahoma, OK 73179
www.climatemaster.com
(405) 745-6000
> *Certification:* Energy Star
> *Warranty:* 10 years on compressor and refrigerant circuit parts;

5 years on other parts; 5-year service labor allowance compressor and refrigerant circuit parts; optional extended warranty. Savings calculator.

Eagle Mountain
4376 Bristol Valley Rd
Canadaigua, NY 14424
www.eagle-mt.com
(800) 572-7831
> Integrates wind, solar, geothermal. Turnkey installation kits.
> *Certification:* ISO 13256, Energy Star
> *Warranty:* 5 years on parts and labor for refrigerant circuit; 2 years on other parts and labor.

 EarthLinked Technologies, Inc
4151 S. Pipkin Rd
Lakeland, FL 33811
www.earthlinked.com
(863) 701-0096
> Direct Exchange (DX) systems, full demand hot water.
> *Certification*: AHRI 870, Energy Star
> *Warranty*: 5 years on parts and labor for compressor parts and optional cathodic installation, 2 years on other parts and labor, 20 years on parts and 5 years on labor of earth loops.

ECR Industries, Inc
PO Box 6469
Reading, PA 19610
www.advgeo.com
(610) 736-0570
> Direct Exchange (DX) systems, on-demand hot water.
> *Certification*: AHRI 870, Energy Star
> *Warranty*: 5 years on compressor, 1 year on other parts, no labor.

ETA
(DX) (Earth to Air Systems)
123 SE Parkway Court
Franklin, TN 37064
www.earthtoair.com
(615) 595-2888
Direct Exchange (DX) systems.
Certification: CSA 222, AHRI 870
Warranty: 5 years, plus extended warranty available.

FHP Manufacturing
601 NW 65th Court
Ft. Lauderdale, FL 33309
www.fhp-mfg.com/residential
(954) 776-5471
Company purchased by Bosch of Germany in 2007. Savings calculator.
Certification: ISO 13256, Energy Star
Warranty: 5 years on compressor, 1 year other parts.

Free Source Energies LLC
(DX) PO Box 122
Redwood Falls, MN 56283
www.freesourceenergies.com
(507) 644-2340
Certification: Energy Star
Warranty: 10 years

GeoComfort
(Enertech Manufacturing)
11699 N. Monique Rd
Northport, MI 49670
www.geocomfort.com
(888) 436-3783
Website savings calculator, GeoSmart financing offered, GeoAnalyst software.
Certification: C13256, ISO 13256, Energy Star
Warranty: 10 years on compressor, 5 years on other parts, extended warranty available.

GeoFurnace Manufacturing
605 4th St. SE
De Smet, SD 57231
www.geofurnacemfg.com
(605) 854-9205
IGSHPA training and accreditation, on demand hot water.
Certification: Energy Star
Warranty: 10 years on refrigerant system.

GeoMaster, LLC
3512 Cavalier Court
Ft. Wayne, IN 46808
www.geoexcel.com
(260) 484-4433
Energy analysis and sizing software.
Certification: ISO 13256, Energy Star
Warranty: 10 years on refrigerant, circuit parts; 5 years on blower parts; 1 year on other parts/labor.

GeoSystems, LLC
Brands: Econar, HydroHeat
7550 Meridian Circle
Maple Grove, MN 55369
www.geosystemsghp.com
(800) 432-6627
Certification: ISO 13256, Energy Star
Warranty: 5 years on parts/labor; extended 10-year warranty available.

Heat Controller, Inc.
1900 Wellworth Ave
Jackson, MI 49203
www.heatcontroller.com
(517) 787-2100
Certification: ISO 13256, Energy Star
Warranty: 12 years for compressor, 6 years on other parts, no labor.

Hydro Delta Corporation
1000 Rico Rd
Monroeville, PA 15146
www.hydroheat.com
(412) 373-5800
   Contractor software: module selection, loop analysis, cost analysis; "On-demand" hot water integrated into geothermal system.
   Certification: C13256, ISO 13256, Energy Star
   Warranty: 5 years on compressor, 1 year on parts; extended 5 & 10 year warranty available.

Hydron Module
(Enertech Manufacturing)
41659 256th St
Mitchel, SD 57301
www.hydronmodule.com
(800) 720-1724
   Savings calculator, GeoAnalyst software.
   Certification: C13256, ISO 13256, Energy Star
   Warranty: 10 years on sealed systems, 5 years on electrical & labor.

Hydro Temp Corporation
PO Box 566
3636 Hwy 67 S
Pocahontas, AR 72455
www.hydro-temp.com
(800) 382-3113
   Priority on-demand hot water system available; kWh meter standard.
   Certification: CSA C13256
   Warranty: 5 years on compressor, 1 year on other parts.

Klimaire, Inc
7909 NW 54th St.
Miami, FL 32166
www.klimaire.com
(305) 593-8358
   Offering advanced infinitely variable compressor.
   Certification: ISO 9002 XXX
   Warranty: 5 years on compressor, 1 year on other parts, no labor.

Sub Terra Energy Systems
24315 NE Dayton Ave.
Newberg, OR 97132
www.subterraenergysystems.com
(503) 487-6326
   Direct Exchange (DX) systems, including installation only in Oregon and SW Washington.
   Certification: Energy Star
   Warranty: 10 years on parts & labor.

TETCO Geothermal
(Enertech Manufacturing)
2506 S. Elm St
Greenville, IL 62246
www.tetcogeo.com
(618) 669-9011
   Savings calculator, GeoSmart financing offered.
   Certification: ISO 13256, Energy Star
   Warranty: 5 years on major parts, 2 years on other parts and labor.

WaterFurnace International
9000 Conservation Way
Fort Wayne, IN 46809
www.waterfurnace.com
(800) 436-7283
   Savings calculator, Geolink sizing software.
   Certification: Energy Star, CSA C13256
   Warranty: 10 years on parts plus labor allowance, 5 years on unit accessories.

Canadian Residential Heat Pump Manufacturers

Boreal Geothermal, Inc.
785 Amherst St
Montreal, Quebec J2L 2J5
www.boreal-geothermal.ca
(450) 534-0203
  Certification: CSA C13256, Energy Star
  Warranty: 5 years replace or repair all internal components.

Enertran Technology, Inc. (ETI)
5 Commerce Road
Orangeville, Ontario L9W 3X5
www.enertran.ca
(800) 941-0053
  On demand hot water, plus systems to eliminate odor and humidity in very air tight homes.
  Certification: CSA C13256
  Warranty: 5 years on compressor, installer must warrant installation for 10 months.

Geofinity Manufacturing
19050 25th Ave
Surrey, BC V3S 3V2
www.geofinitymanufacturing.com
(604) 536-4544
  Certification; ISO 13256, Energy Star
  Warranty: 10 years

Geoflex Systems, Inc.
1069 Clarke Rd
London, Ontario N5V 3B3
www.geoflexsytems.com
(519) 488-1653
  Certification: CSA C13256
  Warranty: 5 years on compressor, 2 years other parts; labor allowance.

GeoSmart Energy
290 Pinebush Road
Cambridge, Ontario N1T 1Z6
www.geosmartenergy.com
(519) 624-0400
  GeoSmart Energy Academy provides training.
  Certification: ISO 13256, Energy Star
  Warranty: 10 years on parts.

Maritime Geothermal Ltd (DX)
PO Box 2555
Petitcodiac, NB E4Z 6H4
www.nordicghp.com
(506) 756-8135
  Nordic line of heat pumps, includes Direct Exchange (DX) systems.
  Certification: CSA C222, AHRI 870, Energy Star
  Warranty: 5 years on all internal parts, labor allowance, extended warranty available.

Northern Heat Pump
3-201 South Railway Ave
Winkler, Manitoba R6W 1JB
www.northernheatpump.com
(204) 325-9772
  ROI calculator on website.
  Certification: ISO 13256, CSA C13256, Energy Star
  Warranty: 5 years, 10 years optional.

PolarBear Geothermal Systems, Inc.
18 Crown Steel Dr
Markham, Ontario L3R 9X8
www.polarbearwshp.com
(888) 485-5854
  Certification: ISO 13256
  Warranty: 5 years on parts/labor.

Appendix D

State Energy Offices

Alabama Energy Division
Call: (334) 242-5290
Fax: (334) 242-0552
www.adeca.alabama.gov/Energy

Alaska Energy Authority
Call: (907) 771-3000
Toll Free in AK: (888) 300-8534
Fax: (907) 771-3044
www.aidea.org/aea

Arizona Energy Office
Call: (602) 771-1137
Fax: (602) 771-1203
www.azcommerce.com/Energy

Arkansas Energy Office
Call: (800) 558-2633
www.arkansasenergy.org

California Energy Commission Renewable Energy Programs
Call: (916) 654-4058
Toll Free in CA (800) 555-7794
Fax: (916) 654-4420
www.energy.ca.gov/renewables

Colorado Energy Office
Recharge Colorado
Call: (303) 866-2100
Toll Free: (800) 632-6662
Fax: (303) 866-2930
www.rechargecolorado.com

Connecticut Energy Office
Call: (860) 418-6200
Toll Free: (800) 286-2214
Fax: (860) 418-6487
www.ct.gov/opm

Delaware Energy Office
Call: (302) 735-3480
Fax: (302) 739-1840
www.dnrec.delaware.gov/energy

District of Columbia Energy Office
Call: (202) 673-6700
Fax: (202) 673-6725
www.ddoe.dc.gov

Florida Energy & Climate Commission
Call: (850) 487-3800
Fax: (850) 922-9701
www.dep.state.fl.us/energy

Georgia Environmental Finance Authority – Energy Initiatives
Call: (404) 584-1000
Fax: (404) 584-1069
www.gefa.org

Hawaii Energy Office
Call: (808) 587-3807
Fax: (808) 586-2536
www.hawaii.gov/dbedt/info/energy

Idaho Office of Energy Resources
Call: (208) 332-1660
Fax: (208) 332-1661
www.energy.idaho.gov

Illinois Bureau of Energy & Recycling
Call: (217) 785-3416
Fax: (217) 785-2618
www.ilbiz.biz/dceo/Bureaus/Energy_Recycling

Indiana Office of Energy Development
Call: (317) 232-8939
Fax: (317) 233-6887
www.in.gov/oed

Iowa Office of Energy Independence
Call: (515) 725-0431
Fax: (515) 725-0438
www.energy.iowa.gov

Kansas Energy Office
Call: (785) 271-3100
Fax: (785) 271-3354
www.kcc.state.ks.us/energy

Kentucky Dept. for Energy Development and Independence
Call: (502) 564-7192
Toll-free in KY: (800) 282-0868
Fax: (502) 564-7406
www.energy.ky.gov

Louisiana Energy Office
Call: (225) 342-1399
Fax: (225) 342-1397
www.dnr.louisiana.gov
(Look under Energy)

Efficiency Maine
Call: (866) 376-2463
Fax: (207) 287-1039
www.efficiencymaine.com

Maryland Energy Office
Call: (410) 260-7655
Toll Free: (800) 72-ENERGY
Fax: (410) 974-2250
www.energy.maryland.gov

Massachusetts Department of Energy Resources
Call: (617) 626-7300
Fax: (617) 727-0030
www.magnet.state.ma.us/doer

Michigan Energy Office
Call: (517) 241-6228
Fax: (517) 241-6229
www.michigan.gov/dleg
(Look under Inside DELEG / Energy Systems)

Minnesota Office of Energy Security
Call: (651) 296-5175
Toll free in MN: (800) 657-3710
Fax: (651) 297-7891
www.energy.mn.gov

Mississippi Energy Division
Call: (601) 359-6600
Toll free: (800) 222-8311
Fax: (601) 359-6642
www.mississippi.org
(Look under Energy)

Missouri Division of Energy
Call: (573) 751-3443
Toll free: (800) 361-4827
Fax: (573) 751-6860
www.dnr.mo.gov/energy

Montana Energy Office
Energize Montana
Call: (406) 841-5200
Fax: (406) 841-5222
www.deq.mt.gov/energy

Nebraska Energy Office
Call: (402) 471-2867
Toll Free: (877) 337-3463
Fax: (402) 471-3064
www.neo.ne.gov

Nevada State Office of Energy
Call: (775) 687-1850
Fax: (775) 687-1869
www.energy.state.nv.us

New Hampshire Office of Energy and Planning
Call: (603) 271-2155
Fax: (603) 271-2615
www.nh.gov/oep

New Jersey's Clean Energy Program
Call: (609) 777-3300
Toll Free: (866) 657-6278
Fax: (609) 777-3330
www.njcleanenergy.com

New Mexico Energy Conservation and Management Division
Call: (505) 476-3310
Fax: (505) 476-3322
www.emnrd.state.nm.us/ecmd

New York State Energy Research and Development Authority
Call: (518) 862-1090
Toll free: (866) NYSERDA
Fax: (518) 862-1091
www.nyserda.org

North Carolina State Energy Office
Call: (919) 733-2230
Toll free in NC: (800) 662-7131
Fax: (919) 733-2953
www.energync.net

North Dakota Office of Renewable Energy & Energy Efficiency
Call: (701) 328-5300
Fax: (701) 328-2308
www.communityservices.nd.gov/energy

Ohio Energy Resources Division
Call: (614) 466-6797
Fax: (614) 466-1864
www.development.ohio.gov/Energy

Oklahoma State Energy Office
Call: (405) 815-6552
Toll free: (800) 879-6552
Fax: (405) 605-2807
www.okcommerce.gov/State-Energy-Office/SEO-Overview

Oregon Department of Energy
Call: (503) 378-4040
Toll free: (800) 221-8035
Fax: (503) 373-7806
www.oregon.gov/energy

Pennsylvania Office of Energy & Technology Deployment
Call: (717) 783-0540
Fax: (717) 783-0546
www.depweb.state.pa.us/energy

Rhode Island Office of Energy Resources
Call: (401) 574-9100
Fax: (401) 574-9125
www.energy.ri.gov

South Carolina Energy Office
Call: (803) 737-8030
Toll free: (800) 851-8899
Fax: (803) 737-9846
www.energy.sc.gov

South Dakota Energy Management Office
Call: (605) 773-3899
Fax: (605) 773-5980
www.state.sd.us/boa/ose/OSE_Statewide_Energy.htm

Tennessee Office of Energy Policy
Call: (615) 741-2994
Fax: (615) 741-0607
www.tn.gov/ecd/CD_office_energy_policy.html

Texas State Energy Conservation Office
Call: (512) 463-1931
Fax: (512) 475-2569
www.seco.cpa.state.tx.us

Utah State Energy Program
Call: (801) 537-3300
Fax: (801) 537-3400
www.geology.utah.gov/sep

Vermont State Energy Office
Call: (802) 828-2811
Toll free: (800) 622-4496
Fax: (802) 828-2342
www.publicservice.vermont.gov
(Look under Renewables & Efficiency)

Virginia Division of Energy
Call: (804) 692-3200
Fax: (804) 692-3237
www.dmme.virginia.gov/divisionenergy

Washington State Energy Policy Division
Call: (360) 725-3118
Fax: (360) 586-8440
www.cted.wa.gov
(Look under Energy Policy)

West Virginia Division of Energy
Call: (304) 558-2234
Toll free: (800) 982-3386
Fax: (304) 558-0362
www.wvcommerce.org/energy

Wisconsin Office of Energy Independence
Call: (608) 261-6609
Fax: (608) 261-8427
www.energyindependence.wi.gov

Wyoming State Energy Program
Call: (307) 777-2800
Toll free: (800) 262-3425
Fax: (307) 777-2838
www.wyomingbusiness.org/business/energy.aspx

Appendix E

Geothermal Careers: Joining the Industry

Jobs Looking for People

If you are considering a career change, you should look carefully at the geothermal field, especially if you have some HVAC experience. It is growing field with a shortage of qualified personnel. High demand often means higher pay.

Most of these jobs are considered to be mid-level or middle-skill careers that demand more than a high-school diploma but less than a four-year degree. These well-paying jobs require training and usually accreditation.

Middle-skill jobs, which are critically important to both the US and Canadian economies, have been disappearing for the last 30 years because of technology changes and off-shoring. Compounding the loss of jobs in this sector was the recent recession.

One bright spot, however, has been the growth of green jobs. "Green" jobs are defined in part, but by no means completely, as jobs in the solar, wind and other renewable energy industries. These have grown 2.5 times faster than all other jobs *(Pew Charitable Trusts report, 2009)*. And in 2010, President Obama established stimulus initiatives to create 5 million new jobs, with an emphasis on that green zone.

As a subset within that green zone, GHP manufacturing shipments are still increasing every year, aided in great part by large tax incentives. Every single GHP shipment from the manufacturer requires a number of local, trained experts in loop construction or borehole drilling, someone to apply software programs to determine system size and ground loop design, someone who understands the GHP circuitry and controls and can hook it all together to make it work. That could be you.

How to Join the Industry

Throughout this book, Michael and I have pointed out that no GHP system can be entirely successful without a high quality installation by trained professionals.

The two career jobs needed most to do this are GHP System Installers and GHP Vertical Loop Installers. For a taste of the type of material covered by the training, note the simplified listing of courses on the next page.

Both of these jobs require US or Canadian certified training plus

GHP System Installer Training Subjects
- IGSHPA Accredited -

Day 1
- Introduction to geothermal industry
- Economics
- Soil identification
- Selecting & sizing the heat pump

Day 2
- Designing the ground-source heat exchanger
- Pipe fusion
- Installation
- Drilling & grouting

Day 3
- Flushing & purging the system
- Heat pump start up
- Open book exam

Courtesy of Michael Hunt, Instructor and CEO of GeoFurnace Manufacturing. For a more detailed course description go to www.geofurnacemfg.com/training

GHP Vertical Loop Installer Subjects
- IGSHPA Accredited -

- GHP system design and layout basics
- System materials
- Pressure drop calculations
- Thermal conductivity
- Drilling processes
- Containment procedures
- Grouting concepts
- Air and debris purging
- Pipe joining techniques
- Project bidding
- Partnerships

Courtesy of IGSHPA

---

field experience. In most cases, the training and certification must come first. Where can you find a friendly, certified trainer in your area? There is no single database that lists all US and Canadian trainers or training organizations. But one comes close. The International Ground Source Heat Pump Association (IGSHPA) lists all IGSHPA accredited trainers in the US and Canada, their organizations, locations and contact information (*www.igsha.okstate.edu/directory*). Or search the internet for "training geothermal heat pumps North Dakota" or whatever state you are in.

You will find many organizations in the US and Canada provide GHP training and certification. It is impractical to list them all here, but I have provided a few samples to show the wide types of organizations that provide GHP training. These include dedicated training companies, GHP contractor/installer companies, GHP manufacturers, and very importantly, non-profit, industry-sponsored geothermal organizations.

I have put these at the top of the list because they provide education and training as well as promote industry standards and set the level of professionalism for the industry.

I am particularly impressed with Canada's approach to professionalism in the GHP field. In Canada, training alone will not provide a certificate. Accreditation is not given until documented experience has also been demonstrated. That is a big difference that will benefit of the homeowner, the industry and the installers (although not all of them may agree with me). In the US, a 3-day workshop provides an installer's card and certificate; job experience is up to the individual or company.

Industry-Sponsored Organizations Providing Training

IGSHPA
www.igshpa.okstate.edu
The International Ground Source Heat Pump Association, located in Stillwater, Oklahoma, is a non-profit, member-driven organization established in 1987 to advance geothermal heat pump technology at local, state, national and international levels. They conduct training on the campus of Oklahoma State University. Traveling workshops are also available. Their annual 3-day conference brings together experts from the US and Canada to share the latest information on ground-source systems. They also produce publications such as *Ground Source Installation Standards*. Successful completion of the training and the exam for the Accredited Installer Workshop or the Accredited Vertical Loop Installer course provides an IGSHPA installer's card, a certificate, a complete set of manuals and membership in the IGSHPA.

Canadian GeoExchange Coalition
www.geoexchange.ca
CGC is an industry-based, non-profit organization located in Montreal, Quebec, Canada. It provides qualification of firms, certification of products, plus training and accreditation of individuals. It offers 3-day workshops for installers and drillers which provides a CGC Training Certificate. To actually receive CGC installer accreditation, the applicant must also demonstrate field experience.

GeoExchange BC
www.geoexchangebc.ca
A non-profit industry association located in Surrey, British Columbia, Canada, that provides information, education and training for the heat pump industry. They offer a 2-year comprehensive Certified Geothermal Technician apprenticeship training program that provides

a Certificate of Qualification. For those with 10 or more years of experience in the industry, a transition period is available to earn a Certificate of Qualification without additional training. In addition, GeoExchangeBC offers CGC-approved training workshops and certification for drillers and installers. Again, a CGC Training Certificate is provided for the application for CGC Installer Accreditation. To actually receive this accreditation, the applicant must also demonstrate field experience.

Iowa Geothermal Association
www.iowageothermal.com
A non-profit, industry-sponsored trade organization in Johnston, Iowa, with a goal of promoting and ensuring quality heat pump installations. They work with the Iowa Energy Center, a research and education organization, to provide IGSHPA GHP installer training and accreditation courses.

Examples of Specialized Training Companies

HeatSpring Learning Institute
www.heatspring.com
An education company located in Cambridge, Massachusetts, that focused on clean energy training. In addition to providing solar energy training, Heatspring offers IGSHPA geothermal installer training as well as a 6-week online entry-level, non-IGSHPA training course.

CleanEdison
www.cleanedison.com
Clean Edison offers education in green building design, solar and wind training, as well as accredited geothermal heat pump training.

Examples of GHP Contractors & Manufacturers Providing Training

GeoFurnace Manufacturing
www.geofurnacemfg.com
De Smet, South Dakota

GeoSmart Energy
www.geosmartenergy.com
Cambridge, Ontario, Canada

Major Geothermal
www.majorgeothermal.om
Wheat Ridge, Colorado

Nexus Energy Products
www.nexusenergyproducts.com
Morden, Manitoba, Canada

Index

"i" represents illustrations or photos

advantages / disadvantages, 71-73
AHRI (Air Conditioning, Heating & Refrigeration Institute, 138
air conditioning (AC), 30, 37, 37-39, 69, 70, 72, 73, 80, 83, 87, 99, 105, 106, 108, 109, 111, 129
air duct system, 47, 51, 131; see also air conditioning
air-source heat pumps, 30, 87
ANSI (American National Standards Institute), 139
antifreeze, 46, 53, 59. *See also* refrigerant
area of influence (heat loops), 112
average annual ground temperatures, 42

backup systems, 78, 90
Btu (British thermal unit), 22, 23, 95, 122-129
building efficiency, 155-156

careers (GHP), 181-184
closed-loop systems, 50, 53-60
   direct exchange , 62-65, 63i, 65i
   horizontal boreholes, 57, 59
   horizontal trenches, 54, 54i
   pond/lake, 53
   slinky coils, 56, 56i, 72i
   vertical, 59, 60i, 60-61, 118
coal, 19-20. *See also* fossil fuels
combustion, 17-18, 79
comfort level, 79-80
commercial/institutional systems, 146-152
   agriculture/horticulture, 147, 152
   airports, 150
   bridges, 148
   case studies, 149, 151
   churches, 150
   gas stations, 148
   government buildings, 148, 152

commercial systems *continued*
   hospitals, 150
   hotels, 148
   schools/colleges, 149-150
comparison of ratings vs. real systems, 128
compressor, 47-48, 63-64, 78-79, 81, 82, 84, 85, 87-88, 112, 113, 126, 134-135, 154-155
constant pressure controller, 83-84
contractors
   finding, 111-114
   qualifications, 112-116
   questions to ask, 113-116
   warranties, 112-116
cooling, home. *See* GHP, cooling
COP (Coefficient of Performance), 103, 113, 124-129, 137, 138, 140
copper tubing, 62-64
cost of operation, 21-25, 22i, 23i
costs, 95-100
   breakdown by segments, 97
   comparisons, 22, 96-100
   DX systems, 99
   online calculators, 95
   rated by tons, 95
CSA (Canadian Standards Association), 139

Database of State Incentives for Renewable Energy (DSIRE), 90, 102
desuperheater, 61, 72, 85, 88, 97
DGX. *See* DX systems
direct exchange systems. *See* DX
direct financing, 153
direct-use geothermal, 30, 147
domestic hot water, 58, 60-61, 85-88
driveway snow melting, 88
ductwork in home, 89, 98

DX (direct exchange) systems, 58, 62-65, 63i, 65i, 66, 112, 149
    2-loop concept, 63i
    area of influence, 112
    attributes, 64-65
    boreholes, 65i
    costing, 99
    early problems, 64
    ground acidity, 65

earth as heat sink, 41
EER (Energy Efficiency Ratio), 103, 127-129, 137, 138
efficiencies of heat pumps, 24, 43, 52, 62, 63, 72, 113, 121, 124, 127, 128, 137-138, 140, 153
efficiency formulas, 124-125
efficiency ratings, 103, 113, 127-129, 136-139
electric baseboard heat, 20-21
electricity, sources of, 21i
    geothermal power, 31-32
energy, definition, 122-123
    transfer of, 131-135
Energy Star guidelines, 101, 112, 137-138

federal tax incentives, 72, 79, 91
finding a contractor, 111-117
fossil fuels, 17, 18-21
frequently asked questions, 164-169
fuel oil, 19-20. *See also* fossil fuels
future trends, 153-156
    direct financing, 153
    geothermal leasing, 153
    increasing efficiencies, 153, 155-156
    official installation standards, 156
    on-board telemetry, 154
    shortage of installers, 156
    solid-state compressors, 154
    variable-speed compressors, 154

geoexchange. *See* GHP

GEP (geothermal electric power), 31-32
Geothermal Heat Pump Consortium, 102
geothermal leasing, 153
geothermal reservoirs, 32
GHP (geothermal heat pumps)
    advantages, 72-73
    comfort level, 79-80
    complexity of install, 73
    See also compressors
    cooling, 47, 48-49, 49i
    costs, 23-25, 95-100
    cost effective, 17
    definition, 15-16, 30-31
    disadvantages, 73
    emissions free, 18
    factors in design, 50
    fuel storage (none), 18
    heating, 43, 46-48, 48i
    how they work, 41-49, 121-129, 131-135
    installation costs, 72, 96-99
    long life, 72
    low cost of operation, 16
    maintenance, 72, 77-78
    operating (electrical) costs, 83-88
    payback, 105-109
    quiet operation, 72, 79
    single system for heating & cooling, 49
    shipments of, 26, 38, 144-145
    sizing the system, 95-99, 117
    superior attributes, 16, 30, 158
    terminology, 26
    See also tons
global marketplace, 143-145
Got Sun? Go Solar, 87, 90
green buildings, 155-156
ground acidity (DX systems), 65
ground-source heat pump.
    See GHP
ground-source options
    factors affecting, 50
    open vs closed loops, 50-60
ground temperatures, 41-42, 42i
growth of GHP market, 144-145

Index

heat exchanger
  domestic water, 85, 87
  GHP, 47-49, 131, 134, 137, 140
  refrigerator, 45
heat pump, definition, 42. *See also* GHP
heat transfer to refrigerant loop, 126-127, 133-135
heating cost comparison data, 22-23
heating domestic water, 85-88
HeatSpring Learning Institute, 88, 98
heat recovery ventilator, 61
home heating. *See* GHP
home's market value, 37, 74-76
homeowner's insurance, 70
homeowners' profiles, 33-40, 55, 58, 60-61, 66, 93-94, 118
horizontal borehole drilling, 57, 59
horizontal closed-loop trenches, 54, 54i
horizontal directional drilling (HDD), 57, 59, 100
HydroHeat, 87
hydronic coil, 47

Ideal Gas Equation, 123, 134
increasing efficiency of GHPs, 153
installation costs. *See* costs
installation standards, 112, 136-139, 156
installation, professional, 65, 111-116
Institute of Social & Economic Research (ISER), 29
insurance, homeowners, 70
ISO (International Standards Institute), 137

Lake Region Electric Cooperative (LREC), 100
Laws of Thermodynamics, 122, 132, 134
leasing, 153
LEED (Leadership in Energy & Environmental Design), 155
lever analogy, 130
loop length, 52

loop systems
  three loops (standard), 46-48
  two loops (DX systems), 62-65, 63i, 65i
lowest cost of operation, 21-25

maintenance, 77-78
Manual J software, 58
manufacturers (heat pumps), 106, 110, 112, 153, 172-176
market value of home with GHP, 37, 74-76
multi-stage compressor, 84

National Home Builders Association, 156
natural gas (NG), 19. *See also* fossil fuels
Natural Resources Canada, Office of Energy Efficiency (OEE), 139
net metering, 90

on-board telemetry, 154
on-demand water heating, 87-88
open-loop systems, 39-40, 47, 50-52, 52i, 67-70, 115
  costs, 97-98
  electricity needed, 67, 70
  See also GHP
  water quality concerns, 51, 68
operating costs, 83-88

payback, 105-109, 116
performance standards of heat pumps, 136-139
photovoltaic. *See* solar electricity
plastic pipes, safety of, 69
pond/lake closed loops, 53
propane (LP or LPG), 19. *See also* fossil fuels
pump-&-dump systems, 51. *See also* open loop systems

radiant heating with GHP, 47, 51, 55, 61, 131
radioactive decay of earth, 41
rebates & tax credits, 101-104
reducing operating costs, 83-88

refrigerant, 43, 47, 62, 64-65, 84, 85, 123, 126, 135
    definition, 126
    in refrigerator, 45, 47
refrigerant loop, 64, 95, 112-113, 126-127, 133-134, 137
refrigerator technology, 15, 42-45
renewable energy
    earth as source, 16-18, 132
    solar electricity, 89-92
    solar water heating, 86, 104
Return on Investment (ROI), 105-109

science & technology of heat pumps, 121-135
    efficiency ratings, 127-129
    heat transfer, 126-127
    how heat pumps deliver heat, 131-135
    refrigerant loop, 126-127
    super efficiencies, 123-126
    water-to-air system, 131
    water-to-water system, 131
savings calculators, 110
scroll compressor, 84
shortage of qualified installers, 156
sizing GHP system, 95-99, 117
    ground loop, 99, 115
slinky coils, 53, 53i, 56, 56i, 72i
software, 58, 100, 113, 114, 117
solar arrays, 90-91, 90i, 91i
solar electricity (photovoltaic), with geothermal, 89-92
    costs & leasing, 91-92
    off-grid homes, 92
    tax credits, 104
solar energy in earth, 16, 17
solar water heating, 86, 104
solid-state compressors, 154
sources of electricity (US & Canada), 21i
standards, 112, 136-139, 156
state energy offices, 177-180
Sub Terra Energy Systems, 100, 112

super efficiencies of heat pumps, 123-126
superior aspects of GHPs, 16-25
swimming pool heating, 87, 104

tankless, on-demand water heating, 86
tax credits, 101-104
    summary, 103-104
thermodynamics, laws of, 122, 132, 134
thermostats, 79, 80-81, 81i
ton (size rating of heat pump), 51, 53, 54, 56, 59, 85, 95-100, 115, 128
Total Green Geothermal, 64, 66, 149
transfer heat from earth, 16
TXV (thermostatic expansion valve), 45, 47, 135

US Green Building Council, 155
utility companies, leasing geothermal loops, 100, 153

variable-speed compressors, 154
vertical closed-loop systems, 59, 60i, 60-61, 94. *See also* GHP

warranties, 65, 72, 112-114
water heating, 58, 60-61, 85-88
water quality concerns for open-loop systems, 51, 68, 70
water well as heat source. *See* open-loop system
water-to-air system, 47, 51, 55, 61, 131
water-to-water system, 47, 51, 55, 61, 131
well source (for GHP). *See* open-loop systems

Zero Energy Home (ZEH), 155-156

List of Graphics

A-1	Geothermal Heat Pump Systems Compared to Fossil Fuel-Based Systems	17
A-2	Type of Heating in Occupied Housing Units	19
A-3	Sources of Electricity (US & Canada)	21
A-4	Heating Cost Comparison Data	22
A-5	Fuel Cost Comparison of Various Heating Sources	23
A-6	Projected GHP Cost Advantage	24
3-1	Average Annual Ground Temperatures in North America	42
3-2	How a Refrigerator Works	44
3-3	3-Loop Transfer of Energy for GHPs	46
3-4	Geothermal Heating in Winter	48
3-5	Geothermal Cooling in Summer	49
4-1	Open-Loop Installation	52
4-2	Closed-Loop Horizontal Installation	54
4-3	Closed-Loop Slinky Coil Installation	56
4-4	Closed-Loop Vertical Installation	60
5-1	2-Loop Transfer of Energy for DX Systems	63
5-2	DX Borehole Installation	65
9-1	Digital Thermostat Schematic	81
10-1	Relative Cost of Heating Domestic Water	86
12-1	Geothermal 3-Ton System Cost Comparison	98
16-1	Geothermal Super Efficiency	125
16-2	Comparison of Energy Star Standards vs. Real Systems	128
16-3	Mechanical Advantage Analogy: The Lever	130
18-1	Energy Star Criteria (Tier 1 & 2)	138
19-1	European Market of Geothermal Heat Pumps	144
19-2	Growth of GHPs in the US Market	145

Acknowledgments

There is a popular belief that a writer as an author creates a book. That is not true. An author creates a manuscript. That manuscript is the heart and soul of a book, but it is not a book.

It requires an experienced editor, especially in a non-fiction genre, to pull together the thousands of words, photos, graphs, sidebars, and paragraphs and then to paginate them in a cohesive way that is pleasing to the eye, has a logical flow, meets literary standards and is actually readable. An index must be created, possibly a glossary, inconsistencies eliminated, and a check for possible copyright infringement. Then, when printed and not before, does it become a book. Of course, at this point the author always calls it his or her book.

A geothermal heat pump is a sort of metaphor for a manuscript. The heat pump manufacturer pulls together past knowledge and standards, then designs and creates a piece of hardware—a heat pump. That is not a residential heating system although it is the heart of that system. It takes an experienced contractor to put together the various pieces—the ground loop, the wiring, the plumbing, the controls, and a room heat delivery subsystem in a cohesive way that has a logical flow, and is actually workable. Then, when the thermostat is turned on, and heat enters the room, and only then, does it become a residential heating system.

But LaVonne Ewing, as my publisher and editor, carried this process one step further. She was familiar with the subject and she contributed interesting ideas, a rejection of bad ideas and offered suggestions during the manuscript creation process. My end was to convert these as well as my own ideas into meaningful, if not eloquent, words and integrate them into the story I am trying to tell. There was a continual, almost twice-a-week exchange of emails brimming with status reports, reactions, etc. This was live-wire stuff, energizing. The collaboration resulted in a far better book, in my opinion.

I have not spent my working life in the geothermal heat pump business. I do have some experience as a homeowner in installing and living with a GHP. But I had to go way beyond that to adequately tell the GHP story. Almost everyone is aware of the potential of the internet in any research project. But until you actually dig in and start searching, you never really appreciate how powerful it can be. This tool made up in part for my lack of experience.

Acknowledgments

But, in addition, I found a network of people actively working in the field who were exceedingly helpful and gracious in answering questions, contributing information, data and relevant material including reading and commenting on the draft manuscript. I want to especially acknowledge a few.

Michael Hunt tops this list. He emailed me one day saying that: "I had always wanted to write a book about heat pumps but never had the time. How can I help you?" We found a way. His background is included on the next page since it is particularly relevant as a contributing (and unpaid) consultant. His introductory comments to this book are fully appreciated. His suggestions and comments have been embedded throughout.

Chuck Russo, my GHP contractor, is one of those very qualified and experienced installers that the world needs more of. He has installed hundreds of GHP systems in the New York, Massachusetts and Vermont area. He was patient with me as I learned how to deal with this new system. He knew that I had never even seen a GHP until it was carted into my basement. So, he was not too impressed when he learned that I was working on a GHP book. That is until he read my Table of Contents and learned that I had a publisher. That did it—he asked for a copy of my manuscript, reviewed it and made helpful suggestions.

Paul Auerbach and Diana Gaeta of Total Green Geothermal (*www.totalgreenus.com*) in Monroe, New York, have provided data on case study examples from their files and important information on DX geothermal systems in which they specialize. Paul has a DX system in his own home and as a contractor and installer, his company installed 107 DX systems in 2009–10. In addition to GHPs, they are also experts at the larger renewable energy picture, including a building's thermal envelope and solar/wind power.

I would like to also thank those homeowners who permitted me to include their homes here as interesting case studies. Then, of course, my thanks to those companies (in addition to Total Green Geothermal above) who provided other case studies around the continent: Robert Hoffman of Hoffman Brothers Heating and Air Conditioning in St. Louis, MO; Randy Waylett of Northern Heat Pump in Manitoba; and Don Murray of Enertran Heat Pumps in Ontario.

And finally, a special thanks to Will Suckow for his drawings that bring a certain charm and an important cohesiveness to the story.

Author Don Lloyd

Don Lloyd is an electrical engineer now retired from Honeywell after 35 years in marketing, engineering and project management. He has extensive experience in taking complex subjects (such as laser gyroscopes or mosaic infrared detectors) and explaining how they work in terms understood by the general public. He is at his best when giving lectures describing how geothermal heat pumps work and why they are the ideal heating and cooling solution for homeowners. Prior to entering the geothermal world, he operated his own software consulting business as a Microsoft Certified Software Professional. He also served as an officer in the US Army.

Don lives in the Hudson River Valley region of New York State with his wife, artist Martha Lloyd, in their new home where the oil trucks pass by but never stop. "Installing a geothermal heat pump was one of the best decisions I ever made!"

Technical Advisor Michael Hunt

As this book's technical advisor, Michael Hunt contributed real-world experience gained from over 27 years of working in the geothermal heat pump industry. He has designed, consulted, and installed heat pump projects throughout North America, and as an accredited IGSHPA installation trainer, he has introduced proper GHP installation and design techniques to over 700 contractors and engineers. Michael is the founder of GeoFurnace Manufacturing Inc, a heat pump manufacturer in South Dakota. His residential and commercial experience includes heat load analysis, loop field design, conductivity testing and field installations, with system designs that include forced hot air, radiant, snow melt, pool dehumidification, domestic hot water and industrial process water applications. His expertise is often used to resolve heat pump and system imperfections, and he performs independent peer review of other system designs, design validation and commissioning.

Visit **PixyJackPress.com** to view all of our renewable energy titles and to order autographed copies.

PO Box 149
Masonville, CO 80541 USA
www.PixyJackPress.com
info@pixyjackpress.com

Printed on chlorine-free, 100% postconsumer recycled paper

The Smart Guide to
Geothermal